Progress in Probability and Statistics
Vol. 6

Edited by
Peter Huber
Murray Rosenblatt

Birkhäuser
Boston · Basel · Stuttgart

Peter Bloomfield
William L. Steiger

Least Absolute Deviations

Theory, Applications, and Algorithms

1983

Birkhäuser
Boston • Basel • Stuttgart

Authors:

Peter Bloomfield
Department of Statistics
North Carolina State University
Raleigh, N.C. 27650

William L. Steiger
Department of Computer Science
Rutgers University
New Brunswick, N.J. 08903

Library of Congress Cataloging in Publication Data

Bloomfield, Peter, 1946-
Least absolute deviations.

(Progress in probability and statistics ; vol. 6)
Bibliography: p.
Includes indexes.
1. Least absolute deviations (Statistics) 2. Regression analysis. 3. Curve fitting. I. Steiger, William L., 1939- . II. Title. III. Series: Progress in probability and statistics ; v. 6.
QA275.B56 1983 519.5 83-25846
ISBN 0-8176-3157-7

CIP-Kurztitelaufnahme der Deutschen Bibliothek

Bloomfield, Peter:
Least absolute deviations : theory, applications, and algorithms / Peter Bloomfield ; William L. Steiger. - Boston ; Basel ; Stuttgart : Birkhäuser, 1983.
(Progress in probability and statistics ; Vol. 6)
ISBN 3-7643-3157-7

NE: Steiger, William L.:; GT

ISBN 0-8176-3157-7
ISBN 3-7643-3157-7
Printed in USA

9 8 7 6 5 4 3 2 1

To our children

Reuben and Nina Steiger
David and Gareth Bloomfield

and our muses

PREFACE

Least squares is probably the best known method for fitting linear models and by far the most widely used. Surprisingly, the discrete L_1 analogue, least absolute deviations (LAD) seems to have been considered first. Possibly the LAD criterion was forced into the background because of the computational difficulties associated with it.

Recently there has been a resurgence of interest in LAD. It was spurred on by work that has resulted in efficient algorithms for obtaining LAD fits. Another stimulus came from robust statistics. LAD estimates resist undue effects from a few, large errors. Therefore, in addition to being robust, they also make good starting points for other iterative, robust procedures.

The LAD criterion has great utility. LAD fits are optimal for linear regressions where the errors are double exponential. However they also have excellent properties well outside this narrow context. In addition they are useful in other linear situations such as time series and multivariate data analysis. Finally, LAD fitting embodies a set of ideas that is important in linear optimization theory and numerical analysis.

In this monograph we will present a unified treatment of the role of LAD techniques in several domains. Some of the material has appeared in recent journal papers and some of it is new.

This presentation is organized in the following way. There are three parts, one for Theory, one for Applications and one for Algorithms.

Part I consists of the first three chapters. Chapter 1 is a short introduction to LAD curve fitting. It begins by tracing the history of the LAD criterion in fitting linear models. The main points in this development involve algorithms or ideas on which algorithms could be constructed. The key section of the chapter develops – from first principles – some of the properties of the LAD fit itself, especially those describing uniqueness and optimality.

Chapter 2 is devoted to linear regression. The behavior of the LAD estimate is described in a result originally due to Bassett and Koenker. This theorem gives the limiting error distribution for LAD, and is shown, in part, to be a consequence of an earlier, more general result on R-estimators, the trick being the identification of LAD as a particular R-estimator. Next, some of the robustness properties are developed for the LAD regression estimator. Finally a Monte-

Carlo experiment compares the behavior of LAD to least squares and to some Huber M-estimators on a variety of regression models.

Chapter 3 deals with linear time series, specifically stationary autoregressions. The main theorem here gives a rate for the convergence of the LAD estimator to the autoregressive parameters. It is surprising that the rate increases as the process becomes more dispersed, or heavy-tailed. Once again, Monte-Carlo results comparing LAD to LSQ and Huber's M-estimator are given for several autoregressions. These portray the behavior described in the main theorem, and convey a sense of the efficiency of LAD in comparison to the other estimators. They also provide evidence for a conjecture that would extend the convergence rate result.

The next two chapters deal with applications and comprise Part II. Chapter 4 treats additive models for two-way tables. It describes some properties of Tukey's median polish technique, and its relationship to the LAD fit for the table. Recent work of Siegel and Kemperman sheds new light on this subject. Chapter 5 discusses the interpretation of the LAD regression as an estimate of the conditional median of y given x. The requirement that the conditional median be a linear function of x is then weakened to the requirement that it merely be a smooth function. This leads to the introduction of cubic splines as estimates of the conditional median and, by a minor modification, of other conditional quantiles.

The final chapters constitute Part III, dealing with algorithmic considerations. We discuss some particular LAD alogrithms and their computational complexities. The equivalence of LAD fitting to solving bounded, feasible linear programming problems is established and the relationship between LAD algorithms and the simplex method is discussed. We conclude with some recent work on exact, finite, algorithms for robust estimation. These emphasize that not only is LAD a member of larger classes of robust methods, but also that the flavor of the LAD procedures carries across to algorithms for other, general, robust methods.

Each chapter concludes with a section of Notes. These go over points somewhat tangential to the material covered in the chapter. Sometimes we trace evolution of ideas, or try to show how items in the existing literature influenced our presentation. Topics that are original are delineated. In addition some conjectures, possibilities for future work, and interesting open questions are mentioned. We thought it preferable to deal with such issues separately, so as not to interrupt the main flow of the presentation.

Literature on LAD techniques appears mainly within two

separate disciplines, statistics and numerical analysis. We have made a reasonably serious attempt to give a comprehensive union of two, fairly disjoint sets of sources.

Our convention for numbering of equations and the like is self-evident. Within sections it is sequential, theorems, definitions, etc., each with their own sequences. Within chapters, i.j refers to item j of Section i. Otherwise i.j.k. refers to item k in Section j of Chapter i. Finally the symbol ■ will mark the end of a proof.

Many debts have accrued since we began working on this book. The first thanks go to colleagues. Professor Michael Osborne sharpened our understanding of the subject. We also benefitted from talking with Eugene Seneta, Scott Zeger, Shula Gross, David Anderson, and Geoffrey Watson. We are especially grateful to Professor Watson for his generous help with support and facilities. We have received support from Department of Energy grant DE-AC02-81ER10841 to the Statistics Department at Princeton. We thank the departments of Computer Science at Rutgers, Statistics at Princeton, and Statistics at North Carolina State University for the stimulating work environments they have provided. Don Watrous in the systems group of the Laboratory for Computer Science Research at Rutgers was our Scribe guru, par-excellence. Computations were carried out on the Rutgers University LCSR

Dec-20. The manuscript was prepared using SCRIBE. Christine Loungo at Rutgers creatively Scribed the first absolutely final draft of the entire manuscript and then made endless corrections. Monica Selinger typed an early draft at Princeton. Finally, apologies to all who wondered about apparent personality changes in the last few frenzied weeks of manuscript preparation, especially family members and pets.

Table of Contents

1. GENERALITIES

1.1 Introduction

Given n points $(\underline{x}_i, y_i) \in R^{k+1}$, the least absolute deviation (LAD) fitting problem is to find a minimizer, $\hat{\underline{c}} \in R^k$, of the AD distance function

$$
\begin{aligned}
f(\underline{c}) &= \sum_{i=1}^{n} |y_i - \sum_{j=1}^{k} c_j\, x_{ij}| \\
&\equiv \sum_{i=1}^{n} |y_i - \langle \underline{c}, \underline{x}_i \rangle| = \sum_{i=1}^{n} |r_i(\underline{c})| \qquad (1) \\
&\equiv \|\underline{y} - X\underline{c}\|_1 \equiv \|\underline{r}(\underline{c})\|_1
\end{aligned}
$$

where $\underline{y} \in R^n$, $\underline{x}_i$ is the ith row of the n x k matrix X, and $\underline{r}(\underline{c}) = \underline{y} - X\underline{c}$ is the vector of residuals. Every $\underline{c} \in R^k$ defines a hyperplane $\pi_{\underline{c}} = \{(\underline{x}, y) \in R^{k+1}: y = \langle \underline{c}, \underline{x} \rangle\}$. Thus $\hat{\underline{c}}$ determines a hyperplane that best fits the n points in the LAD sense. If the first column of X is composed of 1's, $\hat{c}_1$ is the intercept in an equation relating y to the remaining x's; otherwise, the fit passes through the origin.

In the curve-fitting context, given a function h, n points $t_1,\ldots,t_n \in R$, and k basis functions $\phi_1,\ldots,\phi_k$, write $y_i = h(t_i)$ and $x_{ij} = \phi_j(t_i)$. The minimizer $\hat{\underline{c}}$ defines $P_k = \sum_{j=1}^{k} c_j\phi_j$, the k term approximation that best fits h on the t_i's in the LAD sense; thus $\sum_{i=1}^{n} |h(t_i) - P_k(t_i)|$ is minimal.

This chapter will reveal some of the important properties of f and $\hat{\underline{c}}$. As a start, Section 2 gives a brief historical survey. We use this device to present some of the interesting evolution of the subject and as a backdrop against which we can introduce some of the elementary ideas. Uniqueness and optimality are treated in Section 3, which contains other mathematical material that will be useful in discussing subsequent topics.

1.2 Historical Background

Surprisingly, LAD curve-fitting appears to have greater antiquity than least squares. Legendre published his "Principle of Least Squares" in 1805. But nearly a half-century earlier, sometime between 1755 and 1757, R.J. Boscovitch, one of the most unusual figures in 18th century science, articulated an interesting criterion for fitting a line to $n > 2$ points in the plane [Eisenhart (1961)]. If $(\bar{x},\bar{y})$ is the centroid of the n points, (x_i,y_i), the Boscovitch line chooses c to minimize

$$\sum_{i=1}^{n} |y_i - \bar{y} - c(x_i - \bar{x})|.$$

This is the line that minimizes the LAD criterion among all lines constrained to pass through the mean of the data.

In 1760 Boscovitch outlined a simple geometric algorithm to find c, apparently having had some difficulty over the computational aspects in his original work. In the turbulent year of 1789, Laplace gave an algebraic rendition, still elegant and revealing enough to warrant paraphrasing here.

With no loss of generality imagine $\bar{x} = \bar{y} = 0$ and seek the LAD line through the origin; that is, minimize

$$(1) \qquad f(c) = \sum_{i=1}^{n} |y_i - cx_i| = \sum_{i=1}^{n} |r_i(c)|$$

which is (1.1), with $k = 1$. We may as well assume that $x_i \neq 0$ because $f(c) = \sum |y_i| + \sum' |y_i - cx_i|$, the first sum over points for which $x_i = 0$, the second, over points for which $x_i \neq 0$, and f is smallest when the second sum is.

Now imagine that $y_i/x_i \le y_{i+1}/x_{i+1}$, which can always be arranged by renumbering the points. If we restrict c to the interval $(y_p/x_p, y_{p+1}/x_{p+1})$, f becomes

$$(2) \qquad f(c) = \sum_{i=1}^{p} |x_i| \, (c - y_i/x_i) - \sum_{i=p+1}^{n} |x_i| \, (c - y_i/x_i)$$

Differentiation reveals that

$$(3) \qquad f'(c) = \sum_{i=1}^{p} |x_i| - \sum_{i=p+1}^{n} |x_i|,$$

a constant which can't decrease if p increases. Since $f \ge 0$ is continuous, it must be piecewise linear and with a non-decreasing derivative, as in Figure 1.

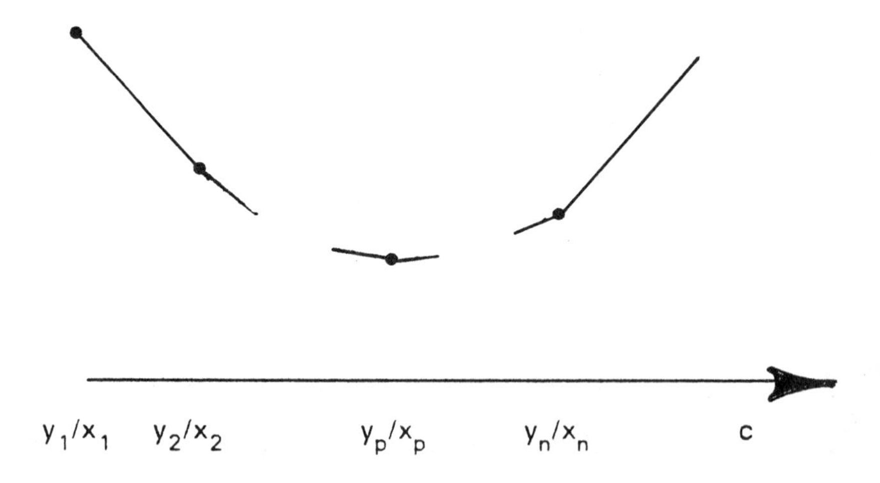

Figure 1. Graph of f

If $f' = 0$ for an interval $J = (y_p/x_p, y_{p+j}/x_{p+j})$, any c in the closure of J minimizes (1). This proves

Lemma 1: f in (1) has a minimizer $\hat{c} = y_i/x_i$ for some $i = 1,\ldots, n$, say $i = p$. Thus, some LAD line through the origin contains (x_p, y_p) so at least one residual in (1), $r_p(\hat{c})$, is zero.

This motivates

Crude Algorithm 1.

1. Compute $c_i = y_i/x_i$, $i = 1,\ldots,n$

2. Evaluate $f(c_i)$, $i = 1,\ldots,n$, and find the minimum, say $f(c_p)$, and minimizer, $\underline{c}_p$.

The computational complexity here is n^2, n division steps to get the c_i and $n-1$ multiplications to evaluate each $f(c_i)$.

Laplace actually gets more from Figure 1. The optimal c is the smallest ratio where the right hand derivative of f is non-negative. From (3) we seek $\min (j: \sum_{i=1}^{j} |x_i| \geq \sum_{i=j+1}^{n} |x_i|)$ or, more simply,

(4) $$p = \min [j: \sum_{i=1}^{j} |x_i| \geq \sum_{i=1}^{n} |x_i|/2].$$

The value $c = y_p/x_p$ is the weighted median of the y_i/x_i with weights $|x_i|$ apparently so named by Edgeworth (1887). His nomenclature becomes clear if you think of a distribution function with steps of size $|x_i|/ \Sigma|x_i|$ at y_i/x_i; its median is c.

The weighted median may be obtained on average in time proportional to nlog(n), n steps to obtain the y_i/x_i and Knlog(n) steps, on the average, to sort the ratios and get the weighted median. More clever approaches, such as only partially sorting the ratios, should improve this bound, probably to a linear one. Combining this with Laplace's argument we obtain

> **Lemma 2:** The LAD line through the origin is a weighted median of y_i/x_i with weights $|x_i|$, (x_i, y_i) being those points for which $x_i \neq 0$. The expected complexity is no more than O(nlog(n)).

Two distinct and interesting problems could arise in this simplest context. First, as alluded to after Figure 1, $\hat{c}$ can be non-unique. For example in fitting (1,1) and (1,2) by a line $y = cx$ through the origin, any $c \in [1,2]$ is optimal. The other problem is degeneracy, where there is more than one point on a LAD fit, as in the problem of fitting (1,1), (2,2), (1,2) by $y = cx$; $y = x$ is optimal but contains 2 data points. Finally, the example (1,1), (2,2), (1,2), (2,4) is optimally fit by $y = cx$ for any $c \in [1,2]$ and exhibits degeneracy at two, of an infinite number of LAD fits.

The foregoing ideas are so central to all that will follow, in theory, applications, and algorithms, that we restate this

SUMMARY: The one parameter LAD fit is a line search for an optimal $c \in R$. It may be characterized as one of n ratios, y_i/x_i, namely the weighted median of the ratios with weights $|x_i|$.

The next important step was taken by Gauss [see Sheynin (1973)]. In 1809 he characterized the LAD fit in R^{k+1} as follows.

<u>**Theorem**</u> <u>1</u>: There is a minimizer $\hat{\underline{c}} \in R^k$ of $f(\underline{c}) = \sum_{i=1}^{n} |r_i(\underline{c})|$ for which $r_i(\hat{\underline{c}}) = 0$ for at least ρ values of i, say $i_1, \ldots, i_\rho$, ρ denoting Rank(X).

In the full rank case this says that the general, k parameter fit is determined by a certain k of the n data points: they have zero residuals and therefore lie in the optimal hyperplane. When $k=1$ it reduces to Lemma 1. It may be the earliest theorem in linear programming and apparently was not deep enough for Gauss to prove. Here is a simple argument showing that if the hyperplane determined by $\underline{c}$ contains fewer than ρ data points, there is a half-line through $\underline{c}$ along which f is piecewise linear, convex, and initially non-increasing; the plane determined by some point $\underline{c}'$ on this line contains at least one more data point than the one determined by $\underline{c}$ and $f(\underline{c}') \leq f(\underline{c})$.

<u>Proof</u>: Given $\underline{c}$, let $\mathbf{Z}_{\underline{c}} = \{i : y_i = \langle\underline{c},\underline{x}_i\rangle\}$ and suppose that $0 \leq |\mathbf{Z}_{\underline{c}}| = p$, where $|\mathbf{A}|$ denotes the cardinality of $\mathbf{A}$. Because $p < \rho$ we can choose $\underline{\delta} \neq \underline{0} \in R^k : \langle\underline{x}_i,\underline{\delta}\rangle = 0,\ i \in \mathbf{Z}_{\underline{c}}$, and $\langle\underline{x}_i,\underline{\delta}\rangle \neq 0$ for some $i \notin \mathbf{Z}_{\underline{c}}$. We then write the line through $\underline{c}$ in the direction $\underline{\delta}$ as

(5) $\underline{c}(t) = \underline{c} + t\underline{\delta},\ t \in R.$

Since $y_i = \langle\underline{c},\underline{x}_i\rangle$ and $\langle\underline{\delta},\underline{x}_i\rangle = 0,\ i \in \mathbf{Z}_{\underline{c}}$,

$$
\begin{aligned}
f(\underline{c}(t)) &= \sum_{i \notin Z_{\underline{c}}} |y_i - \langle\underline{c},\underline{x}_i\rangle - t\,\langle\underline{\delta},\underline{x}_i\rangle| \\
&= \sum_{i \notin Z_{\underline{c}}} |w_i - tz_i|
\end{aligned}
\tag{6}
$$

where $w_i = y_i - \langle\underline{c},\underline{x}_i\rangle = r_i(\underline{c}) \neq 0$ and $z_i = \langle\underline{\delta},\underline{x}_i\rangle$.

Because $z_i \neq 0$ for some $i \notin \mathbf{Z}_{\underline{c}}$, Lemma 2 applies to show that the minimizer of (6) is $t_q = w_q/z_q$, $q \notin \mathbf{Z}_{\underline{c}}$, the weighted median of the w_i/z_i with weights $|z_i|$. Note also that $t_q \neq 0$ since $w_i \neq 0$, $i \notin \mathbf{Z}_{\underline{c}}$.

In passing from the current fit $\underline{c}$ (containing $p < \rho$ data points) to $\underline{c}(t_q)$, the LAD criterion has not increased because $f(\underline{c}) = f(\underline{c}(0)) \geq f(\underline{c}(t_q))$. The new fit contains $p + 1$ points, $(\underline{x}_q, y_q)$ and $(\underline{x}_i, y_i)$, $i \in \mathbf{Z}_{\underline{c}}$. Since the argument applies whenever $p < \rho$, the theorem is proved. (Note that the line search along $\underline{c}(t)$ that located $\underline{c}(t_q)$ may be accomplished by using a weighted median calculation.)■

Assume X has full rank, k. Theorem 1 implies that $\hat{\underline{c}}$ could be computed using the following

Crude Algorithm 2.

For each distinct subset $J = \{j_1,\ldots,j_k\}$ of $\{1,\ldots,n\}$ of size k

(i) when possible solve $y_i = \langle \underline{c}, \underline{x}_i \rangle$, $i \in J$, for $\underline{c}$.
$\hat{\underline{c}}$ is the $\underline{c}$ that minimizes f.

The need to solve $\binom{n}{k}$ linear equation systems of size k is a measure of brute stupidity. All current practical algorithms utilize the Gauss linear programming characterization in Theorem 1, but in a more sensible way. The idea is to pass from an initial fit $\underline{c}_o$ through a sequence $\underline{c}_1,\ldots,\underline{c}_N = \hat{\underline{c}}$ on which f is non-increasing. At every step $j > k$, $\underline{c}_j$ is determined by a certain k data points $i_1,\ldots,i_k$, one of which is replaced in passing from step j to j + 1. While the details properly belong to the discussion of algorithms in Chapter 6 it is still instructive to contemplate the problems that can arise in trying to use the ideas in Lemmas 1 and 2 to obtain higher dimensional fits; it is also appropriate to our historical motif.

In 1887 Edgeworth presented a general procedure which, when k = 2 seeks to fit a line y = a + bx to the n points $\underline{P}_1 = (x_1,y_1),\ldots,\underline{P}_n = (x_n,y_n)$. Take $m_o = 1$ and treat $\underline{P}_{m_o}$ as the origin (subtract $\underline{P}_1$ from the other points). Lemma 1 shows that the fit for $\underline{P}_2,\ldots,\underline{P}_n$, constrained to pass through $\underline{P}_1$, will contain one of these n−1 points, say $\underline{P}_{m_1}$. Now treating $\underline{P}_{m_1}$ as the origin obtain the LAD fit of the other n−1 points obtaining, by Lemma 1, a point $\underline{P}_{m_2}$, etc. Clearly no more than r = n−1 steps − each a weighted median calculation − can be taken before termination, when $\underline{P}_{m_r} = \underline{P}_{m_{r-2}}$. For k > 2 the algorithm is entirely analogous, but more complex [see also Rhodes (1930)].

Edgeworth (1888) expressed surprise that the problem

(1.1), and his algorithm, could permit non-unique solutions, a point which was politely made by Turner (1887). Edgeworth (1888) then showed how to deal with non-uniqueness, an event he regarded as highly unlikely. Actually Edgeworth's algorithm has more grievous defects, due to degeneracy (in the general case, more than k zero residuals at a fit). For example when k = 2 and there are three points on some lines it could cycle through $i_1 \rightarrow i_2 \rightarrow i_3 \rightarrow i_1$, etc., never decreasing f [see Sposito (1976)]. It could also terminate prematurely if, for example, the line through $\underline{P}_{i_1}$ contains point i_2, and vice-versa, but is not optimal [Karst (1968)]. Current implementations can avoid both possibilities [see Sadovski (1974), e.g., for k = 2 and Bartels, Conn, and Sinclair(1978) or Seneta and Steiger (1983) for arbitrary k].

The difficulties illustrated by algorithmic considerations, even in the k = 2 parameter case, point to the need for a careful description of f, of conditions that characterize its minimizers $\hat{\underline{c}}$, and the problems involved in iterating towards it. These issues are addressed in the next section.

1.3 Some Mathematical Background

In this section we outline some of the properties of the function f in (1.1) and the set

$$\mathbf{M} = \{\underline{c} \in R^k : f(\underline{c}) \leq f(\underline{d}) \text{ for all } \underline{d} \in R^k\}$$

of minimizers of f. This is usually done using convex analysis and results from linear programming. The reader will notice our attempt to proceed from first principles.

Let f be given by (2.1). Perhaps the simplest facts are in

Theorem 1: f is continuous and convex.

Proof: Given $\underline{c}$, $\underline{d} \in R^k$ and $t \in [0,1]$,

$$
\begin{aligned}
f(t\underline{c} + (1-t)\underline{d}) &= \sum_{i=1}^{n} |y_i - \langle t\underline{c} + (1-t)\underline{d}, \underline{x}_i\rangle| \\
&= \sum_{i=1}^{n} |t(y_i - \langle\underline{c},\underline{x}_i\rangle) + (1-t)(y_i - \langle\underline{d},\underline{x}_i\rangle| \\
&\le t \sum_{i=1}^{n} |y_i - \langle\underline{c},\underline{x}_i\rangle| + (1-t) \sum_{i=1}^{n} |y_i - \langle\underline{d},\underline{x}_i\rangle| \\
&= tf(\underline{c}) + (1-t)f(\underline{d})
\end{aligned} \tag{1}
$$

the second line following from the linearity of $\langle\cdot,\cdot\rangle$, so f is convex.

Continuity is clear from

$$
\begin{aligned}
&|f(\underline{c}) - f(\underline{d})| \le |\Sigma|y_i - \langle\underline{c},\underline{x}_i\rangle| - \Sigma|\ y_i - \langle\underline{d},\underline{x}_i\rangle|| \\
&\le \Sigma|\langle\underline{c}-\underline{d},\underline{x}_i\rangle|.
\end{aligned} \tag{2}
$$

As $\underline{c} \longrightarrow \underline{d}$, $f(\underline{c}) \longrightarrow f(\underline{d})$.■

The key fact about **M** is that it is non-empty. In the well-behaved cases it is even bounded. As before, let $|\mathbf{A}|$ denote the cardinality of the set **A**.

<u>**Theorem 2**</u>**:** Given f in (1.1), $\mathbf{M} \neq \phi$. If the matrix $X = (\underline{x}_i)$, $\underline{x}_i \in R^k$, has (full) rank $k \le n$, **M** is bounded; otherwise $|\mathbf{M}| > 1$ and **M** is unbounded.

Proof: First suppose rank(X) = k. Then there is $t > 0$ such that for any $\underline{a} \in R^k$ with $||\underline{a}|| = (\sum_{i=1}^{k} a_i^2)^{1/2} = 1$,

$$(3) \qquad \sum_{i=1}^{n} |\langle \underline{a}, \underline{x}_i \rangle| \geq t,$$

since otherwise some $\underline{a} \neq \underline{0} \in R^k$ is orthogonal to all $n \geq k$ of the $\underline{x}_i$.

Let $\underline{c} \in R^k$ be given and write $T = ||\underline{c}||$. By (3), $\sum |\langle \underline{c}, \underline{x}_i \rangle| \geq tT$ so for all $\underline{c} \in R^k$

$$(4) \qquad \begin{aligned} f(\underline{c}) + f(\underline{0}) &= \sum_{i=1}^{n} |y_i - \langle \underline{c}, \underline{x}_i \rangle| + \sum_{i=1}^{n} |y_i| \\ &\geq \sum_{i=1}^{n} |\langle \underline{c}, x_i \rangle| \\ &\geq t\ ||\underline{c}||. \end{aligned}$$

If we now take $\underline{c}$ so $||\underline{c}|| > 2(\sum_{i=1}^{n} |y_i|)/t$ then (4) implies

$$(5) \qquad f(\underline{c}) > f(\underline{0})$$

so no minimizer of f can be outside the compact set

$$\mathbf{S} = \{\underline{c} : ||\underline{c}|| \le 2\sum_{i=1}^{n} |y_i|)/t\}.$$

Hence $\mathbf{M} \subseteq \mathbf{S}$ is bounded. Since f is continuous, it has minima inside $\mathbf{S}$ so $\mathbf{M} \neq \phi$.

If rank (X) = $\rho < k$ choose ρ independent columns $i_1,\dots,i_\rho$ of X, store them in the matrix X_1 and the remaining $k-\rho$ columns in X_2, and write $f_1(\underline{a}) = ||\underline{y} - X_1\underline{a}||$, $\underline{a} \in R^\rho$. The foregoing argument shows there is a bounded $\underline{a} \in R^\rho$ which minimizes f_1.

Write $\underline{c}^T = (\underline{a}^T,\underline{b}^T)$, $\underline{b} \in R^{k-\rho}$, 'T' denoting transpose, and B for the ρ by $k-\rho$ matrix satisfying $X_2 = X_1B$. Then

$$\begin{aligned} f(\underline{c}) &= \underline{y} - X\underline{c} \\ &= \underline{y} - X_1\underline{a} - X_1B\underline{b} \\ &= \underline{y} - X_1(\underline{a}+B\underline{b}) \\ &= f_1(\underline{a}+B\underline{b}) \end{aligned}$$

This shows that $\hat{\underline{c}} \in R^k$: $\hat{c}_{i_j} = \hat{a}_j$, $\hat{c}_i = 0$, $i \neq i_i,\dots,i_\rho$, minimizes f. Thus $\mathbf{M} \neq \phi$ in general. Now however, other $\hat{\underline{c}}$, even unboundedly large ones, will also exist, as long as $B\underline{b}=\underline{0}$. For example in fitting the points (0,0), (0,1), (0,2), by a line y = ax+b, the matrix

$$X = \begin{pmatrix} 1 & 0 \\ 1 & 0 \\ 1 & 0 \end{pmatrix}$$

has rank $1 < k = 2$. Any line $y = 1 + ax$ is optimal so $\underline{c} = (1,a)$ is a LAD fit which may be arbitrarily large.■

The foregoing example also reveals the possibility that $f(\underline{c}_1) = f(\underline{c}_2) \leq f(\underline{c})$ for all $\underline{c} \in R^k$, and $\underline{c}_1 \neq \underline{c}_2$. In fact multiple minima of (1.1) can occur even if X has full rank as can be seen by trying to fit (1,1) and (−1,1) by a line $y = cx$ through the origin. Here $X = \binom{1}{-1}$ has (full) rank 1, yet $f(c) = |1+c| + |1-c| \geq 2 = f(\hat{c})$ as long as $|\hat{c}| \leq 1$.

From now on we shall assume that $\rho = \text{rank}(X) = k$. If this were not the case, the construction used in the proceeding proof would apply to reduce the problem to a new one of (full) rank, ρ.

Given $\underline{c} \in \mathbf{M}$ let $\mathbf{Z}_{\underline{c}} = \{i : r_i(\underline{c}) = 0\}$, $\mathbf{N}_{\underline{c}} = \{i : r_i(\underline{c}) < 0\}$, and $\mathbf{P}_{\underline{c}} = \{i : r_i(\underline{c}) > 0\}$ be the set of indices of zero, negative, and positive residuals, respectively. When there is no ambiguity, the subscript $\underline{c}$ may be dropped. If X has a column of 1's there is certain balance between the number of positive and negative residuals. Specifically,

Theorem 3: If $\underline{c} \in \mathbf{M}$ and $x_{i1} = 1$, $i=1,\ldots,n$, then

(6) $\quad \big|\,|N_{\underline{c}}| - |P_{\underline{c}}|\,\big| \le |Z_{\underline{c}}|$

<u>Proof</u>: Let n_1, n_2, and n_3 denote $|N_{\underline{c}}|, |P_{\underline{c}}|$, and $|Z_{\underline{c}}|$, respectively, so $n = n_1 + n_2 + n_3$. Assume that $n_1 \le n_2$ and write $\delta = \min |r_i(\underline{c})|$, the minimum taken over $N_{\underline{c}} \cup P_{\underline{c}}$.

Given $\epsilon > 0$ let $\underline{d} = (\epsilon, 0, \ldots, 0) + \underline{c}$. If $\epsilon < \delta$, because $x_{i1} = 1$, $r_i(\underline{d}) = r_i(\underline{c}) - \epsilon > 0$ for $i \in P_{\underline{c}}$, $r_i(\underline{d}) = r_i(c) - \epsilon < 0$ for $i \in N_{\underline{c}}$, and $r_i(\underline{d}) = -\epsilon$ for $i \in Z_{\underline{c}}$. Therefore

$$
\begin{aligned}
f(\underline{c}) &= \sum_{i=1}^{n} |r_i(\underline{d})| \\
&= \sum_{P_{\underline{c}}} |r_i(\underline{c})| - n_2\epsilon + \sum_{N_{\underline{c}}} |r_i(c)| + n_1\epsilon + \sum_{Z_{\underline{c}}} \epsilon \\
&= f(\underline{c}) + \epsilon(n_1 - n_2 + n_3)
\end{aligned}
$$

If $n_2 - n_1 > n_3$, $f(\underline{d}) < f(\underline{c})$ contradicting $\underline{c} \in M$. A similar argument can be used if $n_1 > n_2$. ■

When $k = 1$, $r_i(c) = y_i - c$, and (6) describes the median of the y_i's. If n is odd and the y_i's are distinct, $|Z_{\underline{c}}| = 1$. If n is even and c is strictly <u>between</u> the two middle y_i's, $|Z_{\underline{c}}| = 0$; otherwise $|Z_{\underline{c}}| = 1$. In every case, however, (6) holds. The condition that X have a column of 1's is necessary as can be seen by fitting $y = cx$ to (1,1), (2,2), (10,0); the LAD line is $y = 0$ but $|P_c| - |N_d| = 2 > 1 = |Z_c|$.

The next fact about **M** is that it is clearly convex. For if $\underline{c}, \underline{d} \in \mathbf{M}$ and $t \in [0,1]$, $t\underline{c} + (1-t)\underline{d} \in \mathbf{M}$ because

$$f(t\underline{c} + (1-t)\underline{d}) = \sum \left| t_i - <\underline{c},\underline{x}_i> + (1-t)(y_i - <\underline{d},\underline{x}_i>) \right|$$

$$\leq tf(\underline{c}) + (1-t)f(\underline{c})$$

$$= f(\underline{c})$$

and the right hand side is minimal. By theorem 2, $|\mathbf{M}| \geq 1$. By convexity, if $|\mathbf{M}| > 1$, $|\mathbf{M}| = \infty$. Nevertheless distinct elements of **M** cannot differ by much relative to the scale of the data. Specifically,

Theorem 4: Given $\underline{c}, \underline{d} \in \mathbf{M}$, $\underline{c} \neq \underline{d}$,

$$(8) \qquad \mathbf{P}_{\underline{c}} \cap \mathbf{N}_{\underline{d}} = \mathbf{N}_{\underline{c}} \cap \mathbf{P}_{\underline{d}} = \phi.$$

In other words, no point $(\underline{x}_i, y_i)$ is "between" two distinct LAD fits so $r_i(\underline{c})r_i(\underline{d}) < 0$ is impossible.

Proof: First suppose $y = cx$ and $y = dx$ are optimal for $(x_i, y_i) \in R^2$, $c < d$, and that $(y_p - cx_p)(y_p - dx_p) < 0$ for some $p \in \{1,\dots,n\}$. This prohibits $x_p = 0$ which would force $(y_p - cx_p)(y_p - dx_p) \geq 0$.

On the other hand, by the discussion preceding Lemma 2.1, $f' = 0$ on (c,d), which means $\sum_A |x_i| = \sum_B |x_i|$, where $\mathbf{A} = \{i : y_i/x_i \leq c\}$, $\mathbf{B} = \{i : y_i/x_i \leq d\}$. This would force $x_p = 0$ which is not possible. We have thus shown that $p \in \mathbf{P}_c \cap \mathbf{P}_d$ or $p \in \mathbf{N}_c \cap \mathbf{N}_d$, a contradiction.

In the general case suppose $\underline{c} \neq \underline{d}$. Let

$$\underline{c}(t) = \underline{c} + t(\underline{d}-\underline{c}),\ t \in [0,1]$$

and note that $\underline{c}(t) \in \mathbf{M}$ by convexity. Thus

$$f(\underline{c}(t)) = \sum |y_i - \langle \underline{c},\underline{x}_i\rangle - t\langle \underline{d} - \underline{c},\underline{x}_i\rangle|$$

$$= \sum |u_i - tv_i|$$

has minima at $t = 0$ and $t = 1$, where $u_i = y_i - \langle \underline{c},x_i\rangle$ and $v_i = \langle \underline{d}-\underline{c},\underline{x}_i\rangle$. This implies that $u_p(u_p-v_p) \geq 0$ for all p, by the preceding argument. Hence $(y_p-\langle \underline{c},\underline{x}_p\rangle)(y_p-\langle \underline{d},\underline{x}_p\rangle) \geq 0$ for all p and the theorem is proved.∎

Theorem 4 says that any data point will lie on the same side of <u>every</u> optimal hyperplane. If the data are fairly dense (as expected, for example, in large samples of statistical data), then the elements of **M** will be fairly close.

A point $\underline{c} \in R^k$ is extreme if $\{\underline{x}_i, i \in Z_{\underline{c}}\}$ spans R^k, so $|Z_{\underline{c}}| \geq k$. This implies that $\underline{c}$ is uniquely determined by $Z_{\underline{c}}$. If $|Z_{\underline{c}}| > k$, $\underline{c}$ is degenerate. It turns out that **M** is characterized by its extreme points and if $\underline{c}$ is not degenerate, it is an easy matter to decide whether $\underline{c} \in$ **M** and if so, whether it is unique.

First, the nature of **M** is described in the following result, which shows that our use of "extreme" for points of **M** coincides with the usual meaning from convex analysis.

Theorem 5: **M** is the convex hull of the finite set of its extreme points.

Proof: First, if $\underline{c} \neq \underline{d}$ are points of **M**, the convex combination $\underline{c}(t) = \underline{c} + t(\underline{d}-\underline{c})$ is not extreme for any $t \in (0,1)$. For suppose $\underline{c}(t)$ is extreme. Then there is a $p \in Z_{\underline{c}(t)} \cap Z'_{\underline{c}}$ (prime denoting complement), since otherwise $Z_{\underline{c}} \supseteq Z_{\underline{c}(t)}$ and then $\underline{c} = \underline{c}(t)$, because $\underline{c}(t)$ is extreme. However $t r_p(\underline{d}) + (1-t) r_p(\underline{c}) = r_p(\underline{c}(t)) = 0$ implies $r_p(\underline{c})\ r_p(\underline{d}) < 0$ which contradicts Theorem 3. This shows that no extreme $\underline{c} \in$ **M** is a convex combination of points of **M**.

On the other hand if $\underline{c} \in$ **M** is not extreme, then it is a convex combination of points of **M**. Because

$$\dim (\underline{x}_i : i \in \mathbf{Z}_{\underline{c}}) < k,$$

there is $\underline{\delta} \in R^k$ which satisfies $<\underline{\delta}, \underline{x}_i> = 0$ for all $i \in \mathbf{Z}_{\underline{c}}$. Writing $\underline{c}(t) = \underline{c} + t\underline{\delta}$, $t \in R$,

$$\begin{aligned} f(\underline{c}(t)) &= \sum_{i=1}^{n} \left| y_i - <\underline{c}, \underline{x}_i> - t <\underline{\delta}, \underline{x}_i> \right| \\ &= \sum_{i \notin Z_{\underline{c}}} \left| u_i - tv_i \right|. \end{aligned} \tag{9}$$

The proof of Theorem 2.1 guarantees that $\underline{\delta} \neq \underline{0}$ may be chosen so $t_q = u_q/v_q \neq 0$ is a weighted median, and $q \notin \mathbf{Z}_{\underline{c}}$. Because $\underline{c}(0)$ is <u>not</u> extreme in $\underline{c}(t)$, $t \in R$, there is a whole interval $|t| < \epsilon$ for which $c(t) \in \mathbf{M}$. Thus all points of **M** are a convex combination of two points of **M** except for the extreme points.

Finally, there are at most $\binom{n}{k}$ extreme points. This establishes the finiteness assertion, and with it, the theorem.■

Although the nature of **M** is explained by the previous results it is important to be able to decide whether a point $\underline{c} \in R^k$ is a member of **M** and if so, if it is unique.

These questions may be addressed using the directional derivatives of f. For example given $\underline{c}$, assume

$$\lim[f(\underline{c} + t\underline{\delta}) - f(\underline{c})]/t, \; t \downarrow 0$$

exists and is non-negative for all directions $\underline{\delta} \neq \underline{0} \in R^k$. Then $\underline{c} \in \mathbf{M}$ since otherwise there is $\underline{d} \in R^k$, $\underline{d} \neq \underline{c}$ with $f(\underline{c}) > f(\underline{d})$ and, by convexity, the directional derivative in the direction $\underline{\delta} = \underline{d} - \underline{c}$ will be negative. Conversely if $\underline{c} \in \mathbf{M}$ there cannot be a downhill direction. Similar reasoning characterizes $\underline{c} \in \mathbf{M}$ as a unique minimizer if and only if all directional derivatives at $\underline{c}$ are positive.

To get some more detail into these criteria, let $\underline{c} \in R^k$, $\underline{\delta} \neq \underline{0} \in R^k$, and t be given. Then

$$\begin{aligned} f(\underline{c} + t\underline{\delta}) &= \sum|y_i - <\underline{c} + t\underline{\delta},\underline{x}_i>| \\ &= \sum|y_i - <\underline{c},x_i> - t<\underline{\delta},\underline{x}_i>| \qquad (10) \\ &= \sum|r_i - tv_i| \end{aligned}$$

where $r_i = y_i - <\underline{c},\underline{x}_i>$ and $v_i = <\underline{\delta},\underline{x}_i>$. The right hand side of (10) can be written as

$$\sum_V|r_i| + \sum_P|v_i|(r_i/v_i-t) - \sum_N|v_i|(r_i/v_i-t) + |t|\sum_Z|v_i| \qquad (11)$$

where in the spirit of Lemma 2.1, we write $\mathbf{V} = \{i: v_i=0\}$,

P = $\{i \notin \mathbf{V}: (r_i/v_i) > 0\}$, **N** = $\{i \notin \mathbf{V}: (r_i/v_i) < 0\}$, and **Z** = $\{i \notin \mathbf{V}: r_i = 0\}$. It is valid as long as $|t| < \min(|r_i/v_i|, i \notin \mathbf{V})$.

Because of the term $|t|$, the expression in (11) is not differentiable at $t = 0$. However it does have one sided derivatives. The right hand derivative in the direction $\underline{\delta}$, $\lim[f(\underline{c} + t\underline{\delta}) - f(\underline{c})]/t$, $t > 0$, is

$$(12a) \quad f'(\underline{c},\underline{\delta}) = \sum_{N}|v_i| - \sum_{P}|v_i| + \sum_{Z}|v_i|$$

while minus the left hand derivative is

$$(12b) \quad f'(\underline{c},-\underline{\delta}) = \sum_{P}|v_i| - \sum_{N}|v_i| + \sum_{Z}|v_i|$$

This helps characterize the points of **M**.

<u>Theorem</u> <u>6</u>: A point $\underline{c} \in R^k$ is in **M** if and only if for all $\underline{\delta} \neq \underline{0} \in R^k$, $f'(\underline{c},\underline{\delta}) \geq 0$. Thus,

$$(13) \quad \sum_{P}|<\underline{\delta},\underline{x}_i>| - \sum_{N}|<\underline{\delta},\underline{x}_i>| \leq \sum_{Z}|<\underline{\delta},\underline{x}_i>|$$

where **P,N,Z**, refer to the indices i for which $<\underline{\delta},\underline{x}_i> \neq 0$ and $(y_i - <\underline{c},\underline{x}_i>)/<\underline{\delta x}_i>$ is <u>p</u>ositive, <u>n</u>egative, <u>z</u>ero, respectively. It is unique if and only if (13) is a strict inequality for all $\underline{\delta}$.

Writing $v_i = \langle \underline{\delta}, x_i \rangle$ and $r_i = y_i - \langle \underline{c}, x_i \rangle$, (13) holds if and only if 0 is a weighted median of the r_i/v_i with weights $|v_i|$. This is just another way of saying that $t = 0$ gives the optimum point on the line $\underline{c}_t = \underline{c} + t\underline{\delta}$, for all $\underline{\delta}$.

Theorem 6, as it stands, could not be used for deciding whether $\underline{c} \in \mathbf{M}$. The need to check (13) for <u>all</u> $\delta \neq \underline{0} \in R^k$ renders it impractical. However it is the basis for other criteria which are useful in actual situations.

By Theorem 1.1 the search for elements of **M** may be confined to extreme points $\underline{c} \in R^k$. If $\underline{c}$ is not degenerate, it is fairly easy to decide whether or not it belongs to **M**. In this case $|\mathbf{Z}_{\underline{c}}| = \dim(\underline{x}_i,\ i \in \mathbf{Z}_{\underline{c}}) = k$ so each $\underline{x}_j$, $j \notin \mathbf{Z}_{\underline{c}}$ may be expressed as a unique linear combination of the $\underline{x}_i$, $i \in \mathbf{Z}_{\underline{c}}$, as in

$$(14) \qquad \underline{x}_j = \sum_{i \in Z_{\underline{c}}} a_{ji}\, \underline{x}_i, \qquad j \notin \mathbf{Z}_{\underline{c}}.$$

Writing

$$(15) \qquad \lambda_i = \sum_{j \notin Z_{\underline{c}}} a_{ji} \operatorname{sign}(r_j(\underline{c})), \qquad i \in \mathbf{Z}_{\underline{c}},$$

we have

Theorem 7: If $\underline{c} \in R^k$ is a non-degenerate extreme point then $\underline{c} \in \mathbf{M}$ if and only if

$$(16) \qquad |\lambda_i| \leq 1, \quad i \in \mathbf{Z}_{\underline{c}}$$

Proof: Choose $\underline{\delta} \neq \underline{0} \in R^k$. From the middle line in (10),

$$f(\underline{c} + t\underline{\delta}) = \sum_{i=1}^{n} |r_i(\underline{c}) - t <\underline{\delta},\underline{x}_i>|,$$

so that for small enough $t > 0$,

$$(17) \qquad f(\underline{c} + t\underline{\delta}) = t \left[\sum_{i \in Z_{\underline{c}}} | <\underline{\delta},\underline{x}_i>| - \sum_{i \notin Z_{\underline{c}}} <\underline{\delta},\underline{x}_i> \text{sign } (r_i(\underline{c}))\right]$$

This gives, analogous to (12),

$$(18) \qquad f'(\underline{c},\underline{\delta}) = \sum_{i \in Z_{\underline{c}}} |<\underline{\delta},\underline{x}_i>| - \sum_{i \notin Z_{\underline{c}}} <\underline{\delta},\underline{x}_i> \text{sign}(r_i(\underline{c})).$$

Now, writing $\underline{x}_j = \sum_{i \in Z_{\underline{c}}} a_{ji}\underline{x}_i$, $j \notin \mathbf{Z}_{\underline{c}}$, and rearranging terms, we obtain

$$(19) \qquad f'(\underline{c},\underline{\delta}) = \sum_{i \in Z_{\underline{c}}} (|<\underline{\delta},\underline{x}_i>| - \lambda_i <\underline{\delta},\underline{x}_i>)$$

where λ_i is defined in (15).

If (16) holds, each term in the sum in (19) is non-negative. By Theorem 6, $\underline{c} \in \mathbf{M}$.

If (16) fails, $|\lambda_p|>1$ for some $p \in \mathbf{Z}_{\underline{c}}$. Choosing $\underline{\delta}$: $<\underline{\delta},\underline{x}_i> = 0$, $i \neq p$, $i \in \mathbf{Z}_{\underline{c}}$ the expression in (19) is negative (else take $-\underline{\delta}$) so by Theorem 6, $\underline{c} \notin \mathbf{M}$.■

<u>Corollary 1</u>: Let $\underline{c}$ be a non-degenerate extreme point. Then $\underline{c} \in \mathbf{M}$ if and only if

(20) $f'(\underline{c},\underline{\delta}_i) \geq 0$, $i \in \mathbf{Z}_{\underline{c}}$.

for each $\underline{\delta}_i \neq \underline{0} \in R^k$, such that $<\underline{\delta}_i,\underline{x}_j> = 0$, $j \in \mathbf{Z}_{\underline{c}}$, $j \neq i$.

<u>**Proof:**</u> Theorem 6 takes care of the necessity. On the other hand, if (20) holds then putting $\underline{\delta} = \underline{\delta}_j$ in (19) one can infer that $|\lambda_j| \leq 1$, $j \in \mathbf{Z}_{\underline{c}}$ so $\underline{c} \in \mathbf{M}$ by Theorem 7.■

A slight, straightforward modification of these arguments yields the following characterization of uniqueness for non-degenerate, extreme $\underline{c} \in \mathbf{M}$.

<u>Corollary 2</u>: $\underline{c} \in \mathbf{M}$ is unique if, in Theorem 7

(or Corollary 1) the inequality (16) (or (20)) is strict.

Corollary 1 is the modification of Theorem 6 that renders it practical: For non-degenerate, extreme $\underline{c} \in R^k$ one need only check directional derivatives along the k lines determined by the orthogonal complements of $\{\underline{x}_i, \ i \in \mathbf{Z}_{\underline{c}}\}$ from which one of the $\underline{x}_i$ has been deleted. Equivalently $|\lambda_i| \leq 1$ may be checked, $i \in \mathbf{Z}_{\underline{c}}$.

When $|\mathbf{Z}_{\underline{c}}| \equiv \dim(\underline{x}_i, \ i \in \mathbf{Z}_{\underline{c}}) = k$, the $\lambda_i = \sum_i a_{ji}$ are easily determined. The defining relations for the a_{ji} is

$$\underline{x}_j = \sum_{i \in Z_{\underline{c}}} a_{ji} \underline{x}_i, \ j \notin \mathbf{Z}_{\underline{c}},$$

or, in matrix form,

$$N = AB \tag{21}$$

where $N = (\underline{x}_j)$, $j \notin \mathbf{Z}_{\underline{c}}$, (its rows are the $\underline{x}_j$, $j \notin \mathbf{Z}_{\underline{c}}$), $B = \{(\underline{x}_i), \ i \in \mathbf{Z}_{\underline{c}}\}$ and A is the n−k by k matrix of a_{ji}'s to be determined. Thus $A = B^{-1}N$ and the λ_i are easily obtained from (15) and Theorem 7. We have thus characterized **M** in a simple way, both conceptually and computationally.

The situation is more delicate if an extreme $\underline{c} \in R^k$ is degenerate, so $|\mathbf{Z}_{\underline{c}}| > k$. Choose $J \subset \mathbf{Z}_{\underline{c}}$ so that $|J| = k$ and $\{\underline{x}_i,\ i \in J\}$ spans R^k. Then analogous to (14), each $\underline{x}_j$, $j \notin J$ may be expressed uniquely as a linear combination of the $\underline{x}_i$, $i \in J \subset \mathbf{Z}_{\underline{c}}$, as in

$$\underline{x}_j = \sum_{i \in J} a_{ji} \underline{x}_i \ , \ j \notin J. \tag{22}$$

For $i \in J$, define λ_i by (15). From (18) we obtain

$$f'(\underline{\delta}, \underline{c}) = \sum_{j \in Z_{\underline{c}}} |<\underline{\delta}, \underline{x}_j>| - \sum_{j \notin Z_{\underline{c}}} <\underline{\delta}, \underline{x}_j> \text{ sign } (r_j(\underline{c}))''$$

and, using (22) and simplifying, we get

$$\begin{aligned} f'(\underline{c}, \underline{\delta}) &= \sum_{i \in Z_{\underline{c}}} |<\underline{\delta}, x_i>| - \sum_{i \in J} \lambda_i <\underline{\delta}, \underline{x}_i> \\ &= \sum_{i \in Z_{\underline{c}} \setminus J} |<\underline{\delta}, \underline{x}_i>| + \sum_{i \in J} (|<\underline{\delta}, \underline{x}_i>| - \lambda_i <\underline{\delta}, \underline{x}_i>) \end{aligned} \tag{23}$$

It is clear that $|\lambda_i| \leq 1$, $i \in J$ implies that $f'(\underline{c}, \underline{\delta}) \geq 0$ for all $\underline{\delta} \neq \underline{0} \in R^k$. Hence, analogous to the sufficiency of (16) in Theorem 7, we have proved

Theorem 8: An extreme $\underline{c} \in R^k$ is in **M** if, for some $J \subseteq \mathbf{Z}_{\underline{c}}$ where $|J| = k$, and $\{\underline{x}_i,\ i \in J\}$ spans R^k,

$$(24) \qquad |\lambda_i| \le 1,\ i \in J$$

It is unique if (24) is strict.

If degeneracy is allowed, it doesn't seem easy to use Lagrange multiplier conditions, (24), to get a direct analog of the necessity of (16) in Theorem 7. For $p \in J$, take $\underline{\delta}_p \neq \underline{0} \in R^k$ orthogonal to the span of $\{\underline{x}_j,\ j \in I,\ j \neq p\}$. Even if $f'(\underline{c},\underline{\delta}_p)$ were non-negative, (23) would only imply

$$\sum_{i \in Z_{\underline{c}} \setminus J} |\langle \underline{\delta}_p, \underline{x}_i \rangle| + 1 \pm \lambda_p \ge 0$$

from which it does not directly follow that $|\lambda_p| \le 1$. Thus it is not straightforward, using these ideas, to obtain a finite set of directions along which, non-negative directional derivatives assure $\underline{c} \in \mathbf{M}$.

Nevertheless Theorem 8 has some practical value. Given any set $J \subseteq \mathbf{Z}_{\underline{c}}$, $|J| = k$ for which $\{\underline{x}_i,\ i \in J\}$ spans R^k, the a_{ji} satisfying (22) may easily be obtained from (21). Thus the $\lambda_i = \sum_{j \notin Z_{\underline{c}}} a_{ji} \operatorname{sign}(r_j(\underline{c}))$ may be computed and (24) tested.

Let $|\mathbf{Z}_{\underline{c}}| = k+d$, $d \geq 0$ measuring the degree of degeneracy of an extreme $\underline{c} \in R^k$. We proceed to describe a finite set of directions $\pm\, \underline{\delta}_i$ which, if all $f'(\underline{c},\pm\underline{\delta}_i) \geq 0$, will imply $\underline{c} \in \mathbf{M}$. (Necessity of $f'(\underline{c},\underline{\delta}) \geq 0$ at $\underline{c} \in \mathbf{M}$ for any set of directions follows from Theorem 6).

For each $i \in \mathbf{Z}_{\underline{c}}$ define $\pi_i = \{\underline{\delta}: \langle\underline{\delta},\underline{x}_i\rangle = 0\}$, the orthogonal complement of $\underline{x}_i$. We focus on distinct sets J of pointer values for which $\dim(\underline{x}_i, i \in J) = k-1$. Clearly $|J| \geq k-1$, and there are at most

$$\binom{k+d}{k-1} + \ldots + \binom{k+d}{k+d}$$

of them.

For each such J let

$$(25) \qquad \Pi_J = \bigcap_{i \in J} \pi_i$$

and note that $\underline{\delta} \in \Pi_J$ implies $\langle\underline{\delta},\underline{x}_i\rangle = 0$ for all $i \in J$. Becuase $\dim\{\underline{x}_i, i \in J\} = k-1$, Π_J is a 1-dimensional subset of R^k. We call J an edge-set and Π_J an edge direction.

When an extreme $\underline{c} \in R^k$ is non-degenerate, $d = 0$, and $\underline{c}$ has k edges. Corollary 1 says that $\underline{c} \in \mathbf{M}$ if and only if f'

≥ 0 along every edge direction. The same statement is true in general.

> **Theorem 9:** An extreme $\underline{c} \in R^k$ is in **M** if and only if, for all edge sets $J \subseteq \mathbf{Z}_{\underline{c}}$.
>
> (26) $\quad f'(\underline{c},\underline{\delta}) \geq 0$, all $\underline{\delta} \in \Pi_J$.

Proof: For each Π_J choose a specific representative $\underline{\delta}_J \neq \underline{0}$. Any $\underline{\delta} \in \Pi_J$ may now be written as $\underline{\delta} = a\underline{\delta}_J$ for some $a \in R$.

The π_i, $i \in \mathbf{Z}_{\underline{c}}$ partition R^k into intersections of at most 2^m half-spaces, $m = |\mathbf{Z}_{\underline{c}}|$. Any $\underline{\delta} \in R^k$, $\underline{\delta} \notin \pi_i$, $i \in \mathbf{Z}_{\underline{c}}$ is either below π_i if $<\underline{\delta},\underline{x}_i> < 0$, or otherwise, above it.

For $a > 0$, $a\underline{\delta}$ is above or below the same π_i as $\underline{\delta}$. Therefore each of the 2^m patterns of above/below defines the interior of a convex cone comprised of all those $\underline{\delta}$ with the same pattern. Thus each vector $\underline{\epsilon} = (\epsilon_1,\dots,\epsilon_m)$ of $\pm$ 1's describes the interior of one of the convex cones.

The non-interior points of the cone are those $\underline{\delta} \in R^k$ which are elements of some of the π_i (so $<\underline{\delta},\underline{x}_i> = 0$ for these $i \in \mathbf{Z}_{\underline{c}}$) but otherwise have the same pattern $\underline{\epsilon}$ as interior points. Every point in the cone is a positive linear combination of those edge directions that belong to the cone.

Suppose $\underline{\delta} \in R^k$, $i \in \mathbf{Z}_{\underline{c}}$, is given. If $<\underline{\delta},\underline{\delta}_J> < 0$, use $-\delta_J$ as the representative of Π_J. Hence $\underline{\delta}$ is expressible as a linear combination

$$\underline{\delta} = \sum a_J \underline{\delta}_J,$$

where $a_J \geq 0$, the sum extending over all edge sets J for which $\underline{\delta}_J$ is in the cone containing $\underline{\delta}$. From (18) and (26) the directional derivative in the direction $\underline{\delta}$ is $f'(\underline{c},\underline{\delta})$ and equals

$$\begin{aligned}
&\sum_{i \in Z_{\underline{c}}} \left|\sum a_J <\underline{\delta}_J,\underline{x}_i>\right| - \sum_{i \notin Z_{\underline{c}}} \left(\sum a_J <\underline{\delta}_J,\underline{x}_i>\right) \mathrm{sign}(r_i(\underline{c})) \\
(27) \quad &= \sum a_J [\sum_{i \in Z_{\underline{c}}} |<\underline{\delta}_J,\underline{x}_i>| - \sum_{i \notin Z_{\underline{c}}} <\underline{\delta}_J,\underline{x}_i> \mathrm{sign}\,(r_i(\underline{c}))\} \\
&= \sum a_J f'(\underline{c},\underline{\delta}_J) \\
&\geq 0.
\end{aligned}$$

The second line follows because for each $i \in \mathbf{Z}_{\underline{c}}$ if $\underline{\delta}$ is above (below) π_i, so is every $\underline{\delta}_J$ in the cone containing $\underline{\delta}$ and all the $<\underline{\delta}_J,\underline{x}_i>$ are either ≥ 0 or else all < 0.■

Theorem 9 gives practical conditions for optimality that could be used by LAD algorithms. Nevertheless in degenerate cases it could be complicated to actually verify that it is permissible to terminate with a current extreme $\underline{c} \in R^k$. The edge sets J and directions Π_J are required as well as the corresponding directional derivatives. The test for a given J is to show that 0 is the optimal point on the line $\underline{c}(t) = \underline{c} + t\,\underline{\delta}_J$, or equivalently, that 0 is a weighted median of $(y_i - \langle \underline{c}, \underline{x}_i \rangle)/(\underline{\delta}_J, \underline{x}_i)$ with weights $\langle \underline{\delta}_J, \underline{x}_i \rangle$, $i \notin Z$. The details of these considerations will be addressed in Part III (on algorithms) where some of the other ideas in this chapter will also play a role.

1.4 Notes

1. Many alternatives to the term LAD have been used. As a general rule, numerical analysis literature has referred to the problem of minimizing $||\underline{y} - X\underline{c}||$, $\underline{y} \in R^n$, $c \in R^k$, $k \le n$, as L_1, discrete L_1, or ℓ_1 curve-fitting [e.g., Abdelmalek (1971, 1974, 1975), Anderson and Steiger (1982), Armstrong and Godfrey (1979), Barrodale and Roberts (1973, 1974), Bartels and Conn (1978), Bartels, Conn and Sinclair (1978), Osborne and Watson (1971), Robers and Ben-Israel (1969), Spyropoulos, Kiountouzis and Young (1972)]. Within the same literature, "overdetermined equation systems in the L_1 norm", or some quite similar terminology, also appears.

Here is a partial list of the other terms that have been used with some of the references to them: Absolute Deviation Curve-fitting (AD) [Armstrong and Frome (1976), Pfaffenberger and Dinkel (1978), Schlossmacher (1973)]; Least Absolute Error (LAE) [Bassett (1973)]; Least Absolute Residuals Regression (LAR) [Rosenberg and Carlson(1970, 1977)]; Least Absolute Value Regression (LAV) [Armstrong, Elam, and Hultz (1977), Armstrong and Frome (1976, 1977), Barrodale and Roberts (1977), Bartels and Conn (1977), Gentle (1977)]; Least Deviations (LD) [Karst (1958)]; Least Total Deviation (LTD) [Davies(1966)]; Minimum Absolute Deviation Regression (MAD) [Ashar and Wallace (1963), Gallant and Gerig (1974), Harris (1950), Kanter and ..Steiger (1974, 1977), Sharpe (1971)]; Minimum Deviation (MD) [Rhodes(1930)]; Minimum Sum of Absolute Errors (MSAE) [Narula and Wellington(1977a, b, c)]; Sum of Absolute Deviations (SAD); [Rao and Shrinivasan (1962)]; Sum of Absolute Errors (SAE) [Orveson (1969)]; Sum of Absolute Value of Deviations (SAV) [Singleton (1940)] . Perhaps LAV is closest to "least squares". We used LAD because it is more euphonic. Some others who agreed with that name are Armstrong and Frome (1976), Gentle, Kennedy and Sposito (1977), An and Chen (1982), and Gross and Steiger (1979), perhaps for the same reason.

2. There is a brief historical survey of LAD in

Gentle (1977). It touches upon aspects that do not appear in Section 1.

3. The treatment of degeneracy is complicated. Some aspects are discussed in Chapter 7. Sadovski (1974) is aware of the problem; his code attempts to diagnose cycling. Seneta and Steiger (1983) characterize degeneracy, but bypass the problems of dealing with it, once diagnosed.

4. Part of Theorem 2.2 and its proof is due to M.R. Osborne.

5. Theorem 3.3 was stated without proof in Gentle, Kennedy and Sposito (1977). They give an example for the necessity of an intercept term that is somewhat misleading because in it, $\underline{x}_i \neq 0$, all i.

6. Gentle, Kennedy, and Sposito (1977) claim that in fitting $y = a+bx$ to (x_i,y_i), $i=1,\dots,n$, there are at most 4 extreme LAD fits. This is false: $(-2,3/2)$, $(-2,-3/2)$, $(-1,1)$, $(-1,-1)$, $(1,1)$, $(1,-1)$, $(2,3/2)$, $(2,-3/2)$ has 6 extreme fits. As $n \to \infty$ the number of extreme fits, even for $k=2$, is unbounded.

7. Theorem 3.4 was stated without proof in Bloomfield (1982). Theorem 3.5 is usually established via linear programming theory. This direct proof seems to be new. Also, the explicit accounting for degeneracy in Theorem 3.8 seems to be new and Theorem 3.9 is original.

8. The weighted medians may be computed efficiently in several ways. If the ratios are kept in a heap [see, e.g., Knuth (1975)], elements may be removed sequentially and the weights tested until the weighted median is obtained. Chambers' (1971) partial quicksort may be modified to test the balance of weights at the current partition element. Both these methods have average complexity n log(n). If the ratio and corresponding weight sequences are stored as a single sequence of points in the plane using a quad-tree, the partial heapsort technique for removing and testing elements should yield a linear algorithm.

2. LAD IN LINEAR REGRESSION

2.1 Introduction

Let $\underline{Z} = (\underline{X},Y) \in R^{k+1}$ be a random vector whose components obey the linear model

$$
\begin{aligned}
Y &= a_1X_1+\ldots+a_kX_k + U \\
&= \langle \underline{a},\underline{X}\rangle + U
\end{aligned}
\tag{1}
$$

where $\underline{a} \in R^k$ and the random variable U, E(U) = m, are given. If $\underline{X}$ and U are independent $E(U|\underline{X}) = E(U)$ almost surely, and

$$
E(\underline{Y}|\underline{X}) = \langle \underline{a},\underline{X}\rangle + m \quad \text{a.s.} \tag{2}
$$

This means that Y has a linear regression on $\underline{X}$. If and only if m = 0 do the coefficients $\underline{a}$ of the model agree with those of the regression, (2). Finally it is often useful to suppose that in (1), U has a unique median and

$$
\text{median } (U) = 0. \tag{3}
$$

Given a random sample $\underline{z}_i = (\underline{x}_i, y_i)$, $i=1,\dots,n$, the object is to estimate $\underline{a}$.

With a slight modification of (1.1.1) we define the LAD estimator $\hat{\underline{a}}_n$ as a minimizer of

$$(4) \qquad f_n(\underline{c}) \equiv n^{-1}\left(\sum_{i=1}^{n} |y_i - \langle \underline{c}, \underline{x}_i \rangle|\right)$$

Clearly the results of Chapter 1 carry over to this new context.

In this chapter we treat some of the statistical properties of LAD estimators of $\underline{a}$ in (1). The next section establishes the strong consistency of $\hat{\underline{a}}_n$ and shows it to have a limiting normal distribution. Section 3 is devoted to some robustness properties of LAD regression estimators, namely breakdown and influence function, while in Section 4 we examine connections between LAD and M and R-estimators. An interesting fact emerges when LAD is regarded as a member of these classes of robust estimators. Finally Section 5 describes the results of a Monte-Carlo experiment that elucidates some of the sampling properties of $\hat{\underline{a}}_n$, especially as compared to those of some other, familiar estimators.

2.2 Some Statistical Properties of LAD Estimators

Suppose that in (2.1) the errors U follow a double exponential law

$$f(t) = (a/2)e^{-a|t|} ,$$

$a > 0$. The likelihood function for the observed residuals of $(\underline{x}_i, y_i)$ is

$$g(\underline{c}) = (a/2)^n \exp(-a \sum_{i=1}^{n} |y_i - <\underline{c}, \underline{x}_i>|).$$

This implies the familiar but trite fact that LAD is maximum likelihood for these regressions. Here we develop some other properties of $\hat{\underline{a}}_n$ by giving a sequence of easily established facts so that the requirements of each property will be clear. Initially we are able to interpret "sample" rather loosely. Specifically, suppose that $\underline{z}_i = (\underline{x}_i, y_i)$ is a <u>stationary and ergodic</u> sequence, each $\underline{z}_i$ having the law of $\underline{Z} = (\underline{X}, Y)$. The first fact is trivial; it identifies the limit of f_n in (1.4), as $n \to \infty$.

<u>**Lemma 1:**</u> If $\underline{z}_i$ is stationary and ergodic, then for each $\underline{c} \in R^k$,

(1) $f_n(\underline{c}) \longrightarrow E(|Y - <\underline{c},\underline{X}>|) = g(\underline{c})$ a.s.

This follows directly from (1.4) and the ergodic theorem. Also, because $Y - <\underline{c},\underline{X}>$ is convex in $\underline{c}$, it is clear that

Lemma 2: The limit function g in (1) is convex on R^k.

An important property is contained in

Lemma 3: If the components of $\underline{X}$ satisfy

(2) $P(<\underline{c},\underline{X}> = 0) = 1 \Rightarrow \underline{c} = \underline{0}$

and U has unique median zero, then $\underline{a}$ is the unique minimizer of g in (1).

Proof: $g(\underline{c}) = E(|Y - <\underline{c},\underline{X}>|) = E(|U - <\underline{c} - \underline{a},\underline{X}>|)$, by (1.1) and, writing **F** for the sigma-field generated by $\underline{X}$,

(3) $g(\underline{c}) = E\{E(|U - <\underline{c} - \underline{a},\underline{X}>| \,|\mathbf{F})\}$

Therefore g is minimized if

$$E(|U - \langle \underline{c} - \underline{a}, \underline{X} \rangle| \,|\mathbf{F})$$

is. As U is independent of **F** and has unique median zero, $\langle \underline{c} - \underline{a}, \underline{X} \rangle = 0$ on **F**, which, by (2), implies $\underline{c} = \underline{a}$ is the unique minimizer of g.∎

Under the assumptions of Lemma 3, the strong consistency of $\hat{\underline{a}}_n$ may now be asserted.

Theorem 1: Let $\mathbf{Z} = (\underline{X}, Y) \in R^{k+1}$ be integrable and suppose $Y = \langle \underline{a}, \underline{X} \rangle + U$, $\underline{X}$ and U independent, and U having unique median at zero. If (2) holds, then

(4) $\quad \hat{\underline{a}}_n \longrightarrow \underline{a} \quad$ a.s.

Proof: Combining the assumptions and foregoing lemmas we know

i) $f_n(\underline{c}) \longrightarrow g(\underline{c})$ almost surely for each $\underline{c} \in R^k$

ii) f_n has a (not necessarily unique) minimizer $\hat{\underline{a}}_n$

iii) g has unique minimizer $\underline{a}$

iv) g is convex

There are two more facts. First, from Theorem 1.3.1., f_n is convex and continuous, and clearly, g is continuous. Also, the sequence f_n is equicontinuous almost surely because

$$|f_n(\underline{c}) - f_n(\underline{c} + \underline{d})| \le (\sum_{i=1}^{n} |<\underline{d},\underline{x}_i>|)/n$$

$$\le [|d_1| \sum_{i=1}^{n} |x_{i1}| + \ldots + |d_k| \sum_{i=1}^{n} |x_{ik}|]/n;$$

the ergodic theorem bounds the right hand side a.s. by $k(\max|d_i|)(\max E|X_i| + \epsilon)$ for n sufficiently large.

These facts imply that $\hat{\underline{a}}_n \longrightarrow \underline{a}$. For if not, there is $\epsilon > 0$ and a subsequence n_j for which $||\underline{a}_{n_j} - \underline{a}|| \ge \epsilon$. Choose $t < \epsilon$ so that $f_n(\underline{c}) \longrightarrow g(\underline{c})$ on the ball

$$B(\underline{a};t) = \{\underline{c}: ||\underline{c} - \underline{a}|| \le t\}.$$

Write $S(t)$ for the surface of the sphere

$$S(t) = \{\underline{c}: ||\underline{c} - \underline{a}|| = t\}$$

and note that each f_{n_j} attains a minimum at, say, $\underline{c}_{n_j} \in S(t)$. Clearly there is a subsequence m_j of n_j and a limit point $\underline{c}_0$ of $\underline{c}_{n_j}$ such that $\underline{c}_{m_j} \longrightarrow \underline{c}_0$.

Write $\sigma = g(\underline{c}_0) - g(\underline{a})$. By (iii) $\sigma > 0$ By equicontinuity $f_{m_j}(\underline{c}_{m_j}) \longrightarrow g(\underline{c}_0)$ so for m_j sufficiently large,

$$f_{m_j}(\underline{c}_{m_j}) > g(\underline{c}_0) - \sigma/4.$$

By (i), $f_{m_j}(\underline{a}) < g(\underline{a}) + \sigma/4$. Subtracting this from the previous inequality, one obtains

$$f_{m_j}(\underline{c}_{m_j}) - f_{m_j}(\underline{a}) > g(\underline{c}_0) - g(\underline{a}) - \sigma/2 = \sigma/2.$$

Since $\underline{c}_{m_j}$ is a minimizer of f_{m_j} in S(t), and since f_{m_j} is convex, f_{m_j} must attain a minimum in $B(\underline{a},t)$, a contradiction.■

If median (U) = $y \neq 0$, it is easier to discuss the consistency of LAD for models, (1.1), with intercept. Suppose $X_1 = 1$ a.s. Then

$$Y = a_1 + a_2X_2 + \dots + a_kX_k + U, \text{ or,}$$

writing $V = U - y$,

$$Y = (a_1+y) + a_2X_2 + \dots + a_kX_k + V.$$

Since median(V) = 0, Theorem 1 shows that a.s.

$$\underline{a}_n \longrightarrow (a_1+y, a_2, \ldots, a_k)$$

and only the intercept term might not be consistently estimated.

Probably the most important statistical property of the LAD estimator is that it usually has a limiting normal distribution.

__Theorem 2__: Let $Y = \langle \underline{a}, \underline{X} \rangle + U$, X and U independent, the second moment matrix C of $\underline{X}$ positive definite, and Y having a continuous positive density f at 0, the unique median of U. Then if the sample $(\underline{x}_i, y_i)$ is stationary and ergodic martingale differences,

$$(5) \qquad n^{1/2}(\hat{\underline{a}}_n - \underline{a}) \longrightarrow \underline{N}_1 \text{ in distribution,}$$

$\underline{N}_1 \in R^k$ denoting a normal vector with mean zero, and covariance matrix $C^{-1}/(2f(0))^2$.

__Proof__: The idea is to approximate $|t|$ by a uniformly convergent sequence of twice differentiable functions, h_n; the rest depends on a martingale CLT and calculations. For $p > 0$ define

$$h_n(t) = \begin{cases} (1 + (n^p t)^2)/(2n^p), & |t| \le n^{-p} \\ |t|, & |t| > n^{-p} \end{cases}$$

and write

$$(6) \qquad g_n(\underline{c}) = \sum_{i=1}^{n} h_n(r_i(\underline{c}))$$

The h_n converge to $|t|$ pointwise, uniformly in t. Also $g_n - f_n$ converges to 0 almost surely because

$$\begin{aligned} |g_n(\underline{c}) - f_n(\underline{c})| &\le \sum_{i=1}^{n} |h_n(r_i(\underline{c})) - |r_i(\underline{c})|| \\ &\le (2n^p)^{-1} \sum_{i=1}^{n} \kappa(|r_i(\underline{c})| \le n^{-p}) \\ &= (2n^p)^{-1} \sum_{i=1}^{n} \kappa(|u_i - \langle\underline{\delta},\underline{x}_i\rangle| \le n^{-p}) \end{aligned}$$

where κ is the indicator function and we have written

$$\underline{\delta} = \underline{c} - \underline{a}$$

If $p > 1/2$ the Marcinkiewicz strong law for martingales [Loeve (1963) p.242] implies that the right hand side converges to zero. From now on, we assume that $p > 1/2$.

Since $\{g_i\}$ is an equicontinuous sequence of convex functions, a minimizer $\underline{c}_n$ of g_n is strongly consistent under the conditions of Theorem 1, or when the LAD fit $\underline{a}_n$ is. This implies that

$$(7) \qquad \underline{c}_n - \underline{a}_n \longrightarrow 0 \qquad \text{a.s.}$$

In fact, more is true:

$$(8) \qquad n^{1/2}(\underline{c}_n - \underline{a}_n) \longrightarrow 0 \quad \text{in probability.}$$

To see this, compute the partial derivative of g_n at $\underline{c}$ with respect to c_j as

$$(9) \qquad \frac{\partial g_n}{\partial c_j}(\underline{c}) = -\sum_{i=1}^{n} x_{ij}\,[(r_i(\underline{c})/n^{-p})\kappa(A_n(i)) + \text{sign}\,(r_i(\underline{c}))\kappa(A_n'(i))]$$

where

$$A_n(i) = \{|u_i - \langle\underline{\delta},\underline{x}_i\rangle| \leq n^{-p}\},$$

and prime denotes complement. Now take the partial derivatives in (9) with respect to c_m to obtain

$$(10) \qquad \frac{\partial^2 g_n}{\partial c_j \partial c_m}(\underline{c}) = \sum_{i=1}^{n} x_{ij}\ x_{im}\ \kappa\ (A_n(i))/n^{-p}$$

Using (10) to define the entries of the matrix $D_n(\underline{c})$ of second order partial derivatives of g_n at $\underline{c}$, we see that

$$(11) \qquad n^{-1}[D_n(\underline{c}) - 2\sum_{i=1}^{n} f(<\underline{\delta},\underline{x}_i>)\ \underline{x}_i\ \underline{x}_i'] \longrightarrow 0$$

in probability, uniformly in $\underline{\delta}$: For each term in the sum in (10), $E(\kappa\{A_n(i)\}) = F(<\underline{\delta},\underline{x}_i> + n^{-p}) - F(<\underline{\delta},\underline{x}_i> - n^{-p}) = E(\kappa^2\{A_n(i)\})$, so by Tchebycheff's inequality $\kappa\{A_n(i)\}/n^{-p} \longrightarrow 2f(<\underline{\delta},\underline{x}_i>)$ in probability as $n \longrightarrow \infty$, uniformly in $\underline{\delta}$. (Here F and f are the distribution and density functions of U, respectively). The strong consistency of $\underline{c}_n$, (11), and the continuity of f at zero together now imply that

$$n^{-1}[D_n(\underline{c}_n)] \longrightarrow 2f(0)C$$

in probability.

Expanding g_n in a second degree Taylor formula about $\underline{c}_n$ shows that $g_n(\underline{a}_n) - g_n(\underline{c}_n)$ equals

$$(12) \qquad <\underline{a}_n - \underline{c}_n, [\frac{\partial g_n}{\partial \underline{c}}](\underline{c}_n)> + (a_n - c_n)'[D_n(\underline{b}_n)](\underline{a}_n - \underline{c}_n)/2,$$

where $\underline{b}_n$ lies between $\underline{a}_n$ and $\underline{c}_n$. The first term in (12) is zero, by definition of $\underline{c}_n$ and, writing λ_n for the smallest eigenvalue of $[D_n(\underline{b}_n)]/n$, we see from (12) that

$$(13) \qquad n^{-1}(\underline{a}_n - \underline{c}_n)'(\underline{a}_n - \underline{c}_n) \leq 2|g_n(\underline{a}_n) - g_n(\underline{c}_n)|/\lambda_n$$

For n large enough and $w > 0$ small enough $P(\lambda_n \geq w) > 1-\epsilon$; $\underline{b}_n \longrightarrow \underline{a}$ now implies that

$$n^{-1}[D_n(\underline{b}_n)] \longrightarrow 2\ f(0)\ C,$$

which is positive definite. Thus (8) follows from (13) and the equicontinuity of g_n.

From (8) the limiting distributions of $n^{1/2}(\underline{a}_n - \underline{a})$ and $n^{1/2}(\underline{c}_n - \underline{a})$ are identical. To study the latter, expand the partials of g_n with respect to $\underline{c}$ in a first degree Taylor series about $\underline{c} = \underline{c}_n$, so

$$\frac{\partial g_n}{\partial \underline{c}}(\underline{a}) = \frac{\partial g_n}{\partial \underline{c}}(\underline{c}_n) + [D_n(\underline{b}_n)]\ (\underline{a} - \underline{c}_n)$$

for some point $\underline{b}_n$ between $\underline{a}$ and $\underline{c}_n$. The first term on the right vanishes because $\underline{c}_n$ minimizes g_n, and thus

(14) $$n^{1/2}(\underline{a} - \underline{c}_n) = \{n^{-1/2} \frac{\partial g_n}{\partial \underline{c}}(\underline{a})\} \; [n^{-1} D_n(\underline{b}_n)]^{-1}$$

From (9), the expression in curly brackets is

$$n^{-1/2} \sum_{i=1}^{n} \underline{x}_i \; u_i \; \kappa(|u_i| \leq n^{-p})/n^{-p}$$

$$+ \; n^{-1/2} \sum_{i=1}^{n} \underline{x}_i \; \text{sign}\,(u_i)\kappa(|u_i| > n^{-p})$$

The expected values and variances of the components of the vectors in the first sum converge to zero, so by the martingale central limit theorem of Billingsley (1961), that sum $\longrightarrow$ 0 in law.

Med(U) = 0 implies that the vectors in the second sum have means converging to zero and covariances to C, respectively. The same theorem implies a limiting normal distribution for the components and, by the Cramer–Wold device [see Billingsley (1968) p.48] the sum converges to $N(\underline{0},C)$ in law. This result, together with (14) and the fact that $n^{-1} D_n(\underline{b}_n) \longrightarrow 2f(0)C$ in probability, proves that

$$n^{1/2}(c_n - \underline{a}) \longrightarrow N(0,C^{-1})/(2f(0))^2$$

in law, which, in view of (8) establishes the Theorem.■

Theorem 1 asserts the strong consistency of LAD under the condition that ($\underline{X}$,Y) is integrable. We can regard Theorem 2 as the statement that $\hat{\underline{a}}_n \rightarrow \underline{a}$ in probability (with rate n^δ, $\delta < 1/2$). It requires a second moment condition on $\underline{X}$ (namely that $C = E(\underline{X}\ \underline{X}')$ is positive definite) and no moment requirement at all on U (hence Y), as long as the density of U is continuous and positive at 0 = median (U). It seems unlikely that the condition on $\underline{X}$ is necessary for consistency since it also implies the stronger result of asymptotic normality, and one cannot imagine much weakening of the conditions on U.

A slightly less general statement than Theorem 2 was proved by Bassett and Koenker (1978). It should be compared with the analogous limit theorem for the least squares estimator

(15) $$\hat{\underline{c}}_n = (Z'Z)^{-1}Z'\underline{y}$$

where Z is the n by k matrix whose i^{th} row is $\underline{x}_i$, and $\underline{y}$ the vector of y_i's, i=1,...,n. If $(\underline{x}_i, y_i)$ is stationary and ergodic and n large enough, K_n, the covariance matrix of $\hat{\underline{c}}_n$, is

(16) $$K_n = E(\hat{\underline{c}}_n \hat{\underline{c}}_n') = \sigma^2 C^{-1}/n,$$

σ^2 denoting var(U) and C = E($\underline{X}$ $\underline{X}'$) the covariance matrix of the design variables, $\underline{X}$. Finally it is familiar that

$$(17) \qquad n^{1/2}(\hat{\underline{c}}_n - \underline{a}) \xrightarrow{L} \underline{N}_2$$

the least squares analogue of (5), where $\underline{N}_2 \in R^k$ denotes a normal random vector of mean $\underline{0}$ and covariance $\sigma^2 C^{-1}$. Comparing σ^2 with $1/(2f(0))^2$ – the asymptotic variances of the sample mean and sample median of U, respectively – the following, important implication of Theorem 2 emerges:

$$(18) \qquad \underline{N}_1 << \underline{N}_2$$

in the sense of having strictly smaller confidence ellipsoids if the median is a more efficient measure of location for U than the mean.

Section 5 will illustrate this important fact for some cases where the density of U either has a "spike" at 0 or else is heavy-tailed. It will also contrast the consistency behavior of LAD and least-squares. But first we touch on the robustness of LAD estimators in linear regressions and study their properties as M and R-estimators.

2.3 Robustness of LAD: Breakdown and Influence

We have already suggested that LAD fits can resist a few large errors in the data. Since the fit is determined by a certain k of the n data points $(\underline{x}_i, y_i)$, $i=1,...,n$ that lie in the optimal hyperplane, it is little affected by perturbations of other data points. In fact, any fit is completely unaffected by any change in the data where the $\underline{x}$-values remain the same and the y-values change so as to maintain the same signs of residuals. To reveal the nature of this sort of phenomenon in a more precise way, in this section we study the extent to which LAD regression estimators exhibit specific, desireable robustness properties.

Donoho (1982) defines the breakdown point of a multivariate location estimator as follows. Let T be a family of location estimators defined for all sample sizes. Suppose a given set of data $\mathbf{X} = \{\underline{x}_1,...,\underline{x}_n\}$ is contaminated by the inclusion of the dataset $\mathbf{Z} = \{\underline{z}_1,...,\underline{z}_m\}$, where $\underline{x}_i, \underline{z}_i \in R^k$. The estimator T "breaks down at $\mathbf{X}$ for contamination of size m" if

$$\text{(1)} \qquad \sup(\ |T(\mathbf{X}) - T(\mathbf{X} \cup \mathbf{Z})|\ ) = \infty,$$

the sup over all datasets $\mathbf{Z}$ of size m. This means that $T(\mathbf{X} \cup \mathbf{Z})$ can be arbitrarily far from $T(\mathbf{X})$.

Let m^* be the smallest amount of contamination for which T breaks down at **X**. Thus

$$m^* = \min \{m : \sup(|T(\mathbf{X}) - T(\mathbf{X} \cup \mathbf{Z})|) = \infty\},$$

the sup again over all **Z** with $|\mathbf{Z}| = m$. Then the breakdown point of T at **X** is the fraction

$$\epsilon^*(T,\mathbf{X}) = m^*/[n + m^*] \tag{2}$$

The poorest behavior occurs when $\epsilon^* = 1/(n+1)$, that is when adding a single perturbing data point can cause arbitrarily large changes in T.

It is not difficult to extend the notion of breakdown to the regression context. The dataset **X** is the set of n points $\{(\underline{x}_i, y_i) \in R^{k+1}\}$ and T may be regarded as a (family of) regression estimator(s), $T(\mathbf{X}) \in R^k$, defined for all sample sizes $n \geq k$. The data are perturbed by the inclusion of a contaminating dataset $\mathbf{Z} = \{(\underline{x}_i, y_i) \in R^{k+1}\}$ containing m new observations and the breakdown point here is the smallest fraction of contaminated data that can cause arbitrarily large changes in the fit.

Within this framework, LAD is not robust because its

breakdown point is $1/(n+1)$: addition of just <u>one</u> contaminating point can have arbitrarily large effects on the fit. To see this intuitively, it is enough to notice that if just one $(\underline{x},y) \in \mathbf{X}$ has $\underline{x}$ sufficiently far from the origin, then the fit will contain $(\underline{x},y)$. Since $y = <\underline{c},\underline{x}>$ for this point, we can force $||\underline{c}||$ to be arbitrarily large by holding $\underline{x}$ fixed and manipulating y. Therefore, this contaminating data point can alter the fit by as much as desired.

To make this precise, given n points $\{\underline{x}_i,y_i\}$, let F denote the collection of subsets $\mathbf{S}$ of $\{1,\dots,n\}$ satisfying

(i) $|\mathbf{S}| = k$

(ii) $\{\underline{x}_i, \ i \in \mathbf{S}\}$ spans R^k

Also, for $\mathbf{S} \in F$ let $\Delta(\mathbf{S})$ be the set of direction vectors $\underline{\delta}$ satisfying

(iii) $||\underline{\delta}|| = 1$

(iv) $<\underline{\delta},\underline{x}> = 0$ for $k-1$ $\underline{x}_i$'s, $i \in \mathbf{S}$.

Clearly each $\Delta(\mathbf{S})$ also spans R^k. Finally, define

(3) $$t = \min \min(\max[\ |\langle \underline{\delta}, \underline{x} \rangle| : \underline{\delta} \in \Delta(S)] : S \in F),$$

the left-most minimization over $||\underline{x}|| = 1$. Note that $t > 0$ since otherwise there is a unit vector $\underline{x}$ and a $S \in F$ such that $\langle \underline{\delta}, \underline{x} \rangle = 0$ for all $\underline{\delta} \in \Delta(S)$, but this is impossible because $\Delta(S)$ spans R^k. Then we have

Theorem 1: If $||\underline{x}|| > t^{-1} \sum_{i=1}^{n} ||\underline{x}_i||$, then the LAD fit, $\underline{c}$ to

$$\{(\underline{x}_i, y_i),\ 1 \le i \le n\} \cup \{(\underline{x}, y)\}$$

satisfies $y = \langle \underline{c}, \underline{x} \rangle$.

Proof: For some $S \in F$, consider the fit $\underline{c}$ determined by $(\underline{x}_i, y_i)$, $i \in S$. By construction, there is a $\underline{\delta} \in \Delta(S)$ such that

$$\begin{aligned} |\langle \underline{\delta}, \underline{x} \rangle| &\ge t||\underline{x}|| \\ &\ge \sum_{i=1}^{n} ||\underline{x}_i|| \\ &\ge \sum_{i=1}^{n} |\langle \underline{\delta}, \underline{x}_i \rangle|. \end{aligned}$$

Thus, more than half the mass of the distribution with masses proportional to $|\langle\underline{\delta},\underline{x}_i\rangle|$ and $|\langle\underline{\delta},\underline{x}\rangle|$ is associated with $(\underline{x},y)$. From the expression for $f'(\underline{c},\underline{\delta})$ in (1.3.12a) and (1.3.12b), $\underline{c}$ is optimal only if $y = \langle\underline{c},\underline{x}\rangle$.■

It could be argued that it is unrealistically pessimistic to allow the perturbing data point to have such a dominant $\underline{x}$. For example, we might allow contamination only by replicating $\underline{x}$'s already in the dataset. This would usually give a larger break-down point, but even in this case there are examples where it is still $1/(n+1)$.

A second concept that is often used to assess the robustness of a procedure is its influence function [see for instance Hampel (1974) or Huber (1980)]. It is usually defined in terms of the response of an estimator to contamination by an infinitesimal amount of additional data. However this approach does not seem fruitful for LAD fits because it is easily seen that neither adding a data point with infinitesimal weight, nor infinitesimally reducing the weight of an existing data point, will change a unique fit. However either of these changes can eliminate at least part of the nonuniqueness of a nonunique fit. Thus the proportional impact of these infinitesimal changes is either zero or infinite.

The influence of each of the data points, singly or in

combination, can of course be assessed by dropping them and refitting, as discussed for instance by Cook and Weisberg (1982, Chapter 3). However we cannot say much about this procedure in LAD regressions. This contrasts sharply with the situation regarding least squares, for which Cook and Weisberg give explicit expressions for the effect on the fit of dropping data points. Moreover in the LAD context, the possible consequences of dropping a single data point, or some subset of the data, ranges from (1) no effect, to (2), passing from one unique fit to another one, to (3), reducing or increasing the extent of nonuniqueness already present. It seems difficult to descscribe in explicit terms what the actual effect will be.

Nevertheless, we can use infinitesimal perturbations to obtain an approximate idea of influence in large samples. We first extend the notion of LAD fitting from finite data sets (which in the extension, will be represented by their empirical distributions in R^{k+1}), to arbitrary probability distributions, F. Suppose that $\underline{c}(F)$ is a minimizer of

$$\int |y - \langle \underline{c}, \underline{x} \rangle| \, dF(\underline{x}, y)$$

the integral over $(\underline{x}, y) \in R^{k+1}$. Suppose also that we can write

$$dF(\underline{\xi},\eta) = dG(\underline{\xi})\ h(\eta|\underline{\xi})\ d\eta$$

and that the conditional density $h(\eta|\underline{\xi})$ is continuous at $\eta = \langle\underline{c}(F),\underline{\xi}\rangle$, for almost all $\underline{\xi}$ (under the measure dG). Then

$$\underline{c}((1-\epsilon)F + \epsilon\delta_{(\underline{x},y)}) = \underline{c}(F) + \epsilon B^{-1}\underline{x}\ \text{sign}[y-\langle\underline{c}(F),\underline{x}\rangle]/2,$$

where $\delta_{(\underline{x},y)}$ denotes the unit mass at $(\underline{x},y)$ and the matrix

$$B = \int \underline{\xi}\ \underline{\xi}^T\ h(\langle\underline{c}(F),\underline{\xi}\rangle \mid \underline{\xi})\ dG(\underline{\xi}).$$

In particular, if the distribution F describes the linear model in (1.1),

(4) $\qquad Y = \langle\underline{a},\underline{X}\rangle + U,$

where $\underline{X}$ and U are independent and U has a unique median of zero, then $\underline{c}(F) = \underline{a}$ by Lemma 2.3 and

$$h(\eta|\underline{\xi}) = f(\eta - \langle\underline{\xi},\underline{a}\rangle),$$

f denoting the density of U, so that

(5) $$B = f(0) \int \underline{\xi}\, \underline{\xi}^T \, dG(\underline{\xi}) = f(0)\, C.$$

Consequently, the influence of infinitesimal contamination at $(\underline{x},y)$ is

(6) $$f(0)^{-1}\, C^{-1}\, \underline{x}\, \text{sign}(y - \langle \underline{a}, \underline{x} \rangle)/2.$$

The robustness of LAD thus lies in the boundedness of this influence as a function of y for fixed $\underline{x}$, while its lack of robustness lies in the unboundedness of the influence as a function of $\underline{x}$.

The expression for the influence function, (6), involves the density function f. This explains the nature of the difficulty in calculating influence for finite data sets. In large data sets with continuously distributed errors, however, the empirical distribution should be close to the smooth parent distribution. Consequently, the influence curve defined for the parent distribution should give a reasonable approximation to the effect of contaminating the finite data set.

2.4 LAD in M and R-Estimation

LAD is a member of two important classes of robust regression estimators, M and R. There is an interesting interplay here between the particular, and the general. In one direction, LAD algorithms motivate some promising techniques for other M and R estimators, but this topic will be deferred until Chapter 7.

However, in the other direction, general properties of R-estimators obviously hold in the particular case of LAD. Thus the limiting normal distribution for R estimates, established in 1972 by Jaeckel, holds for LAD once it is shown that LAD is an R-estimator and satisfies the conditions of that theorem. One purpose of this section, therefore, is to demonstrate this apparently unknown, though simple fact.

Given a sample $(\underline{x}_i, y_i) \in R^{k+1}$, $i=1,\ldots,n$ of size n and $\underline{c} \in R^k$, write $r_i(\underline{c}) = y_i - \langle \underline{c}, \underline{x}_i \rangle$ for the ith residual. M-estimators arise as the minimizers $\hat{\underline{c}}_n$ of a function h on R^k defined by

$$(1) \qquad h(\underline{c}) = \sum_{i=1}^{n} \rho(r_i(\underline{c})),$$

where the function ρ on R is given. LAD occurs in the special case

$$\rho(t) = |t| .$$

In the growing literature on M–estimators, there exist theorems giving conditions for asymptotic normality, but all seem to require differentiability of ρ. Hence any limiting distributions of LAD estimators must be deduced from another set of ideas.

To this end, suppose we are given a set of n <u>scores</u> $d_1 \leq \ldots \leq d_n$, $d_1 < d_n$, with

$$(2) \qquad \sum_{i=1}^{n} d_i = 0$$

R–estimators arise as the minimizers $\hat{\underline{c}}_n$ of

$$(3) \qquad g_n(\underline{c}) = \sum_{i=1}^{n} d_i r_{(i)}(\underline{c}),$$

$a_{(i)}$ denoting the ith order statistic of $a_1,\ldots,a_n$.

R–estimation makes most sense in models containing intercept terms, even though it does not provide an estimate of the intercept. Suppose that $X_i = 1$ almost surely, and that $\underline{e} = (1,0,\ldots,0) \in R^k$. Then (2) and (3) imply that

$$g_n(\underline{c} + t\underline{e}) = g_n(\underline{c})$$

for all t so the R-criterion does not distinguish between $\underline{c} + t\underline{e}$ and $\underline{c}$. (If the model does not contain an intercept, R-estimation essentially fits it as if a constant term were included, but then ignored.) The following result shows that if the d_i are taken to be sign-scores

$$(4) \qquad d_i = \begin{cases} -1 & \text{if } i < (n+1)/2 \\ 0 & \text{if } i = (n+1)/2 \\ 1 & \text{if } i > (n+1)/2 \end{cases}$$

and if the intercept term is estimated appropriately, then the R-estimator of $\underline{a}$ in

$$(5) \qquad Y = \langle \underline{a}, \underline{X} \rangle + U$$

agrees with the LAD estimator. Specifically

Lemma 1: Let $(\underline{x}_i, y_i) \in R^{k+1}$, $i = 1,\ldots,n$ be a sample of size n from (5) in which $X_1 = 1$ almost surely and let $\gamma \in R^k$ be a minimizer of (3), where the d_i are the sign scores, (4). If

$$\underline{c}(\underline{\gamma}) = (\gamma_1 + \text{med}(r_j(\underline{\gamma})), \gamma_2, \ldots, \gamma_k)$$

then $\underline{c}(\underline{\gamma}) = \hat{\underline{a}}_n$, the LAD estimator of $\underline{a}$.

Proof: $\Sigma|r_i(\underline{c}(\gamma))| = \Sigma|r_i(\underline{\gamma}) - \text{med}\{r_j(\gamma)\}|$

$$= \Sigma\, d_i [r_{(i)}(\underline{\gamma}) - \text{med}\{r_j(\underline{\gamma})\}]$$

$$= \Sigma\, d_i\, r_{(i)}(\underline{\gamma})$$

$$\leq \Sigma\, d_i\, r_{(i)}(\underline{c})$$

$$\leq \Sigma\, |r_{(i)}(\underline{c})|$$

for any $\underline{c} \in R^k$, all sums over $i = 1,\ldots,n$, so $\underline{c}(\underline{\gamma})$ is a minimizer of the AD function. The steps, respectively, are based on: the construction of $\underline{c}(\underline{\gamma})$; the property that half the deviations from a median are positive and half negative; the constraint (2); the assumption that $\underline{\gamma}$ minimizes (3); the triangle inequality.∎

The LAD estimator does not seem to be an R-estimator in models without an intercept (more generally in models for which $\underline{1} = (1,\ldots,1) \in R^n$ is not almost surely in the column space of $\underline{X}$). In this case, to assert the asymptotic normality of $\hat{\underline{a}}$ we must resort to Theorem 2.2. However for the model

$$Y = a_0 + \langle \underline{a}, \underline{X} \rangle + U$$

we appeal to Lemma 1 and the following result, quite similar to one of Jaeckel (1972), omitting a few technical assumptions on $\underline{X}$ and the scores d_i.

Theorem 1: If the density f of U has finite Fisher–Information, C, the covariance matrix of X, is positive definite, $(\underline{x}_i, y_i) \in R^{k+1}$ is an i.i.d. sample from $Y = a_0 + \langle \underline{a}, \underline{X} \rangle + U$, and scores d_i satisfy (2), then the R–estimator $\hat{\underline{c}}_n$ satisfies

$$(6) \qquad n^{1/2}(\hat{\underline{c}}_n - \underline{a}) \xrightarrow{L} \underline{N}_3,$$

$\underline{N}_3 \in R^k$ normal with mean zero and covariance $s^2 C^{-1}/(2f(0))^2$. If the d_i are sign scores, the constant $s = 1$

2.5 Sampling Behavior of LAD

In this section we describe a Monte–Carlo experiment that illustrates the behavior of LAD estimates in linear regression. We consider random samples $(\underline{x}_i, y_i) \in R^3$, $i=1,\ldots,n$ from the linear model

$$(1) \qquad Y_i = a_0 + a_1 X_1 + a_2 X_2 + U$$

for various sample sizes, n, and distributions of U. The results illustrate some of the theory already established, and suggest aspects not yet delineated by theorems.

To hint at asymptotics, we used three sample sizes. In small samples, n = 10 was used. The value, n = 50 was used for moderate sized samples and n = 100, for large samples.

A variety of densities, f, were used for U, among them the standard normal, the double exponential

$$f(t) = (e^{-|t|})/2,$$

and the logistic. In addition some experiments were performed when U is a contaminated normal.

Consider the density

$$(2) \qquad f_\epsilon(t) = \begin{cases} ((1-\epsilon)e^{-t^2/2})/(2\pi), & |t| \le k \\ ((1-\epsilon)e^{-k(|t|-k/2)})/(2\pi), & |t| > k, \end{cases}$$

normal on [−k,k], and exponential outside. If $k(\epsilon)$ is a solution of $2\phi(k)/k - 2\Phi(-k) = \epsilon/(1-\epsilon)$, ϕ and Φ denoting the normal density and distribution, respectively, f_ϵ has a continuous derivative and has minimum Fisher information for location amongst all ϵ contaminated distributions [c.f. Huber (1981)]. In our experiments we took

$$\epsilon = .05 \quad \text{and} \quad k = 1.399 = k(.05)$$
$$\epsilon = .25 \quad \text{and} \quad k = .766 = k(.25)$$

Next we contaminated normals using heavier tails than exponential. Specifically if P_a has density

$$(3) \qquad g_a(t) = a/(2(1+|t|^{1+a})) \ , \ t \in R,$$

it is symmetric Pareto of index a. For $a > 2$ it has finite variance but if $a \le 2$, $E(|P_a|^\beta) < \infty$ iff $\beta < a$. We chose $a = 1.2$ which implies finite mean but infinite variance.

The two heavy tailed distributions we used were h_ϵ,

$\epsilon = .05$ and $\epsilon = .25$ where, analogous to (2), h_ϵ is normal in $[-k(\epsilon),k(\epsilon)]$ and P_a otherwise. These densities have discontinuities at plus and minus $k(\epsilon)$, and they have infinite variance. In fact, as ϵ increases, the Pareto tails have greater probability.

Finally, we did some experiments where U is simply Pareto $P_{1.2}$, as in (3). In summary, the 8 different sampling situations are:

A)	Normal	E)	$h_{.05}$
B)	$f_{.05}$	F)	$h_{.25}$
C)	$f_{.25}$	G)	Pareto(1.2)
D)	double exponential	H)	Logistic

To highlight the behavior of the LAD estimator, $\hat{\underline{a}}_n$, we compared it to several others. The natural comparison is with least squares (2.15). By the remark following (2.18) one expects LAD to be asymptotically superior in sampling situations D, E, F, G. In fact, LAD is maximum likelihood for D while in (E) – (G) the few, large values of U that will occur should disturb LAD much less than least squares.

Another interesting comparison is with certain M-estimators. If, in (4.1) the function ρ is chosen to satisfy

$$(4) \qquad \rho(t) = \begin{cases} t^2/2, & |t| \leq \delta \\ \delta|t| - \delta^2/2, & |t| > \delta \end{cases}$$

the resulting estimator, $\hat{\underline{c}}_\delta$, defined as a minimizer of

$$H_\delta(\underline{c}) = \sum_{i=1}^{n} |\rho_i(\underline{c})|,$$

is called a Huber M–estimator. If δ is large enough ρ treats all residuals quadratically, and $\hat{\underline{c}}_\delta$ is least squares. On the other hand, if δ is small enough, ρ treats all residuals linearly, and $\hat{\underline{c}}_\delta$ is LAD.

These estimators are closely related to the contaminated normals in (2). For each $\epsilon > 0$ a random variable X_ϵ following (2) is normal, given that $|X_\epsilon| \leq k(\epsilon)$ and double exponential, given that $|X_\epsilon| > k(\epsilon)$. If one chooses $\delta = k(\epsilon)$ in (4), $\hat{\underline{c}}_\delta$ is maximum likelihood for $\underline{a}$ in (1.1) if U has density f_ϵ and $\underline{a} = (a_0, \underline{0}) \in R^{k+1}$. Thus, corresponding to $\epsilon = .05$ we choose $\delta = 1.399 = k(.05)$ and we call the corresponding estimator $\hat{\underline{c}}_{.05}$; for $\epsilon = .25$ we take $\delta = .766 = k(.25)$ and call the estimator $\hat{\underline{c}}_{.25}$. Because $k(0) = \infty$ and $k(1) = 0$, it makes sense to call least squares $\hat{\underline{c}}_0$ (minimizes $H_\infty(\underline{c})$ and also max–likelihood for regression with normal errors) and LAD, $\hat{\underline{c}}_1$

(minimizes $H_0(\underline{c})$ and also max-likelihood for regression with double exponential errors) and we shall do so. Huber (1981) calls f_ϵ in (2) the "least informative" density for estimating a_0 with $\hat{\underline{c}}_\epsilon$.

If U is normal, one might expect

(5) $$\hat{\underline{c}}_0 << \hat{\underline{c}}_{.05} << \hat{\underline{c}}_{.25} << \hat{\underline{c}}_1,$$

at least asymptotically, << denoting, roughly, "has a tighter error distribution". Similarly, if U is $f_{.05}$, one could suppose

(6) $$\hat{\underline{c}}_{.05} << \hat{\underline{c}}_\beta, \ \beta \neq .05$$

but it is not clear how least squares, LAD, and $\hat{\underline{c}}_{.25}$ will be ordered, and how this order might depend on sample size. The analogous statement holds for $\hat{\underline{c}}_{.25}$ when U follows the density $f_{.25}$ but here, because contamination is greater, the asymptotic preference ordering might be revealed at smaller sample sizes. Finally, if U is double exponential one expects

(7) $$\hat{\underline{c}}_0 >> \hat{\underline{c}}_{.05} >> \hat{\underline{c}}_{.25} >> \hat{\underline{c}}_1$$

The same ordering would be expected when contamination is by the heavy–tailed Paretos (3), at least for large samples, and also in the pure Pareto situation, G.

The actual experiments were performed as follows: a distribution for U was chosen. Then, on each replication a sample of size n was generated from

$$(8) \qquad Y = 1 + X_1 + X_2 + U.$$

The values of X_1 were always taken to be

$$a_i = [i - (n+1)/2]/[(n^2-1)/12]^{1/2}, \; i=1,\ldots,n.$$

They have mean zero and variance 1. For X_2, n standard exponential random numbers, centered at 0, were generated from the density $f(t) = e^{-(t+1)}$, $-1 \leq t$. Finally, n values for U were generated and $y_i = 1 + a_i + x_{i2} + u_i$ computed, $i=1,\ldots,n$.

For each sample, the estimators $\hat{\underline{c}}_0$, $\hat{\underline{c}}_{.05}$, $\hat{\underline{c}}_{.25}$ and $\hat{\underline{c}}_1$ were computed and the errors from $\underline{a} = (1,1,1)$ recorded. The entire process of sample generation and estimation was repeated 901 times. This large number of replications assures that characteristics of the sampling distribution like quartiles, medians, etc. may be approximated with more than 1 place accuracy.

The detailed results of the study appear in 24 tables in the Appendix. For each sample size, n=10, n=50, and n=100, there are eight tables, one for each distribution of U. In each table the max, min, 5 percent point, median, 1st quartile, 3rd quartile and the 95 percent point are given for the sampling distribution of errors of each of the 4 estimators. We summarize some of this information in the following tables. In Tables 1–3, the 4 by 8 arrays give the mean length of the error vector for each estimator, under each error distribution for U, one table for each sample size.

Tables 1-3 Mean of $\|\hat{\underline{a}}-\underline{a}\|$ for the estimators of $\underline{a}$ in (1) each based on 901 samples of size n.

Table 1 (n=10)

Distributions of U

	A	B	C	D	E	F	G	H
$\hat{\underline{c}}_0$	.615	.637	.787	.845	1.781	5.031	5.326	1.092
$\hat{\underline{c}}_{.05}$	.622	.626	.674	.797	.909	2.790	2.899	1.106
$\hat{\underline{c}}_{.25}$	.644	.660	.627	.791	.879	2.346	1.513	1.165
$\hat{\underline{c}}_1$	.735	.766	.644	.853	.943	2.398	1.514	1.276

Table 2 (n=50)

	A	B	C	D	E	F	G	H
$\hat{\underline{c}}_0$	.231	.240	.310	.331	.754	3.170	7.196	.424
$\hat{\underline{c}}_{.05}$	.238	.228	.226	.290	.290	.701	.366	.413
$\hat{\underline{c}}_{.25}$	.250	.258	.191	.276	.281	.588	.314	.425
$\hat{\underline{c}}_1$	.290	.296	.190	.286	.318	.553	.293	.465

Table 3 (n=100)

	A	B	C	D	E	F	G	H
$\hat{\underline{c}}_0$	.165	.162	.219	.231	.778	3.428	3.338	.291
$\hat{\underline{c}}_{.05}$	.168	.155	.156	.201	.194	.462	.248	.283
$\hat{\underline{c}}_{.25}$	.177	.183	.129	.191	.190	.334	.209	.295
$\hat{\underline{c}}_1$	.207	.213	.128	.185	.218	.307	.181	.326

As expected, the ordering in (5) held for normal regressions. Not surprising it also obtained for logistic errors. The orderings predicted in (6) and (7) also were basically borne out by the data. It is not clear why (7) fails to hold for the heavy tailed error distributions (E) and (F). Perhaps $\hat{\underline{c}}_{.25}$ is actually best in these regressions.

2.6 Notes

1. The requirement in Lemma 2.3 is that for any linear subspace $S \subset R^k$, $P(\underline{X} \in S) = 0$, the inclusion being proper. It is easy to see that this condition is necessary for uniqueness of the minimizer of g.

2. Theorem 2.1 is not the first result describing the strong consistency of LAD. Amemiya(1979) proves it for independent and identically distributed samples when U has infinite mean. The present proof is similar to that of Gross and Steiger (1979) from the context of time series and allows the independence assumption to be relaxed. Notice that i.i.d. samples are not required.

3. The assumptions of Theorem 2 do not imply Bassett-Koenker. Even though $(\underline{x}_i, y_i)$ stationary and ergodic implies that the design sample covariance matrix $(Z'Z)/n$ is positive definite for n large, the non-independence of the $y_i - \langle \underline{c}, \underline{x}_i \rangle$ requires new central limit theory. The idea of the proof is similar to, but simpler than, that of Amemiya (1979) which seems to have some mistakes. Ruppert and Carroll (1977) have taken a related approach for the location problem.

4. LAD is consistent for infinite variance regressions where least squares certainly is not. Suppose U,V are i.i.d. and integrable and write $Z = U + V$, $X = U$. Then X has a linear regression on Z with slope $a = 1/2$ because

$$\begin{aligned} E(X \mid Z) &= E(U \mid U + V) \\ &= E(V \mid U + V) \\ &= E(U + V \mid U + V)/2 \\ &= Z/2 \end{aligned}$$

However if U and V are taken to be symmetric stable random variables of index $\alpha < 2$, the least squares estimator $\hat{c}_n = \Sigma x_i z_i / \Sigma z_i^2$, based on an i.i.d. sample (z_i, x_i), converges in distribution to $S/(S + T)$ S,T being independent, positive stable random variable of index $\alpha/2$ [see Kanter and Steiger (1974)] so least squares is not consistent. A result in Kanter and Steiger (1977) shows that if $\alpha > 1$, the LAD estimator

$\hat{a}_n \to 1/2$ in probability. Perhaps the most convenient source for material on stable laws and their domains of attraction is Feller (1971).

5. Jaeckel (1972) does not point out that choosing the scores as in (4.6) yields the LAD estimator nor do Bassett and Koenker (1978) acknowledge that asymptotic normality of LAD can follow from Jaeckel (1972) for unconstrained regressions. Furthermore, although Hogg (1979) discusses R-estimation for regressions using the sign-scores, he does not mention any connection with LAD regression. Hence it is reasonable to suppose that Lemma 4.1 is new. It was mentioned by M. Osborne in a seminar in Canberra in 1980.

6. Monte-Carlo experiments apparently demonstrate the inconsistency of least squares when X_2 and U both have infinite variance. On the other hand, Kanter and Steiger (1974) have shown that the least squares estimator is consistent if $Y = aX + U$, X and U being i.i.d. random variables attracted to a stable law of index $\alpha \in (0,2)$, hence having infinite variance. This is a different linear model from that described in Note 4.

7. Strong consistency of least squares seems more difficult to establish than for LAD [see Lai, Robbins, and Wei (1978), Lai and Robbins (1980), and Nelson (1980)]. The last two also allow non i.i.d. samples. In contrast, asymptotic normality seems easier prove for least squares than for LAD. We do not know whether this is a fact or an artifact and can offer no explanation.

8. The influence of data points on least squares regression estimates has been discussed by Belsley, Kuh, and Welsch (1980) and Cook and Weisberg (1980, 1982), among others.

9. The influence function for M-estimators of regression coefficients is described in Krasker and Welsch (1982).

3. LAD IN AUTOREGRESSION

3.1 Introduction

This chapter is devoted to stationary, k^{th} order autoregressions

(1) $$X_n = a_0 + a_1 X_{n-1} + \ldots + a_k X_{n-k} + U_n$$

where $\ldots, U_{-1}, U_0, U_1, \ldots$ is an i.i.d. sequence of random variables. The first question that arises is: what conditions on $\underline{a}$ and the U_i will assure that there actually exists a stationary sequence $\{X_i\}$ satisfying (1)? Using (1) recursively for X_{n-1} in (1), then again for X_{n-2}, etc., one obtains after N substitutions,

(2) $$X_n = d(N) + \sum_{j=0}^{N} b_j U_{n-j} + \sum_{j=N+1}^{N+k} c_j(N) X_{n-j},$$

for a certain unique choice of $b_j, c_j(N)$; i.e., $b_0 = 1$, $b_1 = a_1$, etc.

If one imposes the stationarity condition

(3) $a_1 z + \dots + a_k z^k \neq 1, \ |z| \leq 1$

for any complex z, then both b_j and $c_j(N)$ can easily be shown to converge to zero geometrically fast as $j \longrightarrow \infty$; in fact there are positive constants B and C and $\beta, \gamma \in (0,1)$ for which

$$|b_j| \leq B\beta^j \text{ and}$$

$$|c_j(N)| \leq C\gamma^j.$$

We can also show that $d(N) \longrightarrow$ to some limit d as $N \longrightarrow \infty$.

Now suppose that

$$\sum_{j=0}^{N} b_j \, U_{n-j}$$

converges almost surely as $N \longrightarrow \infty$, and let

$$\tilde{X}_n = d + \sum_{j=0}^{\infty} b_j \, U_{n-j}$$

Then $\tilde{X}_n - a_o - a_1 \tilde{X}_{n-1} - \dots - a_k \tilde{X}_{n-k}$ is easily shown to be U_n, so $\{\tilde{X}_n\}$ is a stationary solution of (1).

Conversely suppose $\{X_n^*\}$ is a stationary sequence satisfying (1). Then the second sum in (2) converges to zero almost surely so the first sum, $\sum b_j U_{n-j}$, must converge almost surely. This establishes that under (3), the almost sure convergence of $\sum b_j U_{n-j}$ is necessary and sufficient for the existence of a stationary solution to (1).

The first sum will converge iff $\sum t^j U_{n-j}$ does for some $t \in (0,1)$, a condition equivalent to

$$(4) \qquad \limsup (|U_{n-j}|^{1/j}) < 1/t, \quad \text{a.s.}$$

By the Borel-Cantelli lemma, (4) holds if

$$\sum_j P[|U_{n-j}| \geq ((1+\epsilon)/t)^j] < \infty$$

or, taking logarithms, if

$$\sum_j P[\log_+ |U_{m-j}| \geq j \log ((1 + \epsilon)/t)] < \infty,$$

$\log_+ t$ denoting $\max(0, \log t)$. Using the latter inequality we have proved the substantial part of the following statement.

Lemma 1: If $\underline{a}$ satisfies (3) and if

$$(5) \qquad E(\log_+ |U_i|) < \infty$$

then $\sum_{j=0}^{\infty} b_j U_{n-j}$ converges a.s. and there is an almost surely unique stationary sequence $\{X_n\}$ satisfying (1).

The problem we now address is that of estimating $\underline{a}$ from a sample $X_1,\dots,X_{N+k}$ of size N + k of the stationary autoregression in (1). The LAD estimator in this context, is a minimizer, $\hat{\underline{a}}_N$, of

$$(6) \qquad f_N(\underline{c}) = N^{-1} \sum_{i=1}^{N} |X_{i+k} - (c_0 + c_1 X_{i+k-1} + \dots + c_k X_i)|$$

In the next section we establish some of the statistical properties of $\hat{\underline{a}}_N$. In particular it is shown to be strongly consistent under very mild conditions. A lower bound on the rate of convergence $\hat{\underline{a}}_N \to \underline{a}$ is also established under only slightly more restrictive conditions. Theorems like these must be compared to the analogous results for least squares due to Kanter and Hannan (1977). To set the scene, define the least squares estimator, $\hat{\underline{c}}_N$, as a minimizer of

$$(7) \qquad g_N(\underline{c}) = N^{-1} \sum_{i=1}^{N} [X_{i+k} - (c_0 + c_1 X_{i+k-1} + \dots + c_k X_i)]^2.$$

Suppose the U_i are attracted to a non-normal stable law of index α, a property we abbreviate using the notation

$$U_i \in \text{Dom}(\alpha), \qquad \alpha \in (0,2).$$

This means that both $t^\beta P(U_n \geq t)$ and $t^\beta P(-t \geq U_n)$ converge to 0 or ∞ as $n \to \infty$, depending on whether $\beta < \alpha$ or $\beta > \alpha$, and implies

$$(8) \qquad E(|U_i|^\beta) < \infty$$

if and only if $\beta < \alpha$.

Kanter and Hannan have shown that

$$(9) \qquad N^\delta(\hat{\underline{c}}_N - \underline{a}) \to \underline{0}, \text{ a.s.}$$

if $\delta < 1/\alpha$. This is remarkable because as $\alpha \to 0$, the U_i, hence the X_n, become extremely heavy tailed, yet $\hat{\underline{c}}_N$ converges faster and faster.

The analogue of (9) for LAD appears in Section 2. Section 3 illustrates the sampling properties of LAD and compares them to those of some other Huber M-estimators via Monte-Carlo experiment analogous to the one described in Section 2.5.

3.2 Behavior of LAD in Autoregressions

Given $\underline{a} \in R^{k+1}$ satisfying the stationarity condition, (1.3), and an i.i.d. sequence ..., $U_{-1}, U_0, U_1, \ldots$, with $E(\log_+ |U_i|) < \infty$, we study the unique stationary process $\{X_i\}$ that satisfies

$$(1) \qquad X_n = a_0 + a_1 X_{n-1} + \ldots + a_k X_{n-k} + U_n.$$

Specifically, using N + k observations $X_1, \ldots, X_{N+k}$ the LAD estimator $\hat{\underline{a}}_N$ is a minimizer of

$$(2) \qquad f_N(\underline{c}) = N^{-1}\left(\sum_{i=1}^{N} |X_{i+k} - (c_0 + c_1 X_{i+k-1} + \ldots + c_k X_i)|\right)$$

The first result establishes the strong consistency of LAD.

Theorem 1: If U is non-lattice, $\mathrm{med}(U_i) = t$ is unique, and $E(|U_i|) < \infty$, then $\hat{\underline{a}}_N \longrightarrow (a_0 + t, a_1,\dots,a_k)$ almost surely.

Proof: This closely resembles the proof of Theorem 2.2.1. By the ergodic theorem,

$$f_N(\underline{c}) \longrightarrow E(|X_{N+k} - (c_0 + \dots + c_k X_N)| \equiv f(\underline{c}),$$

a convex function of $\underline{c}$. Since U_i is non-lattice so is X_i so $(X_{N-1},\dots,X_{N-k})$ satisfies Lemma 2.2.3 which establishes $(a_0 + t, a_1,\dots,a_k)$ as the unique minimizer of f. The equicontinuity of f_N follows here from

$$|f_N(\underline{c}) - f_N(\underline{c}+\underline{d})|$$

$$\leq N^{-1}(\sum_{i=1}^{N} |d_0 + d_1 X_{i+k-1} + \dots + d_k X_i|)$$

$$\leq |d_0| + k(\max|d_i|)\, E(|X_i| + \epsilon)$$

the last inequality another consequence of the ergodic theorem. This clinches the almost sure convergence of $\hat{\underline{a}}_N$ to the minimizer of f, as asserted, using the final argument from Theorem 2.2.1.■

It is likely that $\hat{\underline{a}}_N$ is strongly consistent even if $E(|U_i|) = \infty$ (see Section 3) but a proof of such a statement would probably not depend on the ergodic theorem. Actually if we impose slightly more on the U_i, a much stronger result is possible. Specifically, we require the U_i to be attracted to a stable law of index $\alpha \in (1,2)$, a condition that implies the existence of all fractional moments $\beta < \alpha$, and in particular, the mean; i.e., $E(|U_i|^{\beta}) < \infty$ for $\beta < \alpha$.

Suppose $a_0 = 0$ in (1). We have

Theorem 2: If the U_i are i.i.d., and in Dom (α), $\alpha \in (1,2)$ and with unique median zero, then

$$N^{\delta}(\hat{\underline{a}}_N - \underline{a}) \longrightarrow \underline{0} \text{ in probability} \tag{3}$$

for $\delta < 1/\alpha$.

Proof: Write $\underline{\Delta}_N = \hat{\underline{a}}_N - \underline{a}$ and define

$$\underline{a}(t) = \underline{a} + t\underline{\Delta}_N, \ t > 0 \tag{4}$$

a point on the half-line from $\underline{a}$ through $\hat{\underline{a}}_N$. If

(5) $N^{\delta} \; ||\underline{\Delta}_N|| > \epsilon$

there is $t < 1$ such that $\underline{a}(t)$ is between $\underline{a}$ and $\hat{\underline{a}}_N$ and $||t\underline{\Delta}_N|| = \epsilon/N^{\delta}$. Otherwise for some $t \geq 1$, $||t\underline{\Delta}_N|| = \epsilon/N^{\delta}$ and $\hat{\underline{a}}_N$ is between $\underline{a}$ and $\underline{a}(t)$. In both cases $\underline{d}_N = \underline{a}(t) - \underline{a}$ is a random variable with $||\underline{d}_N|| = \epsilon N^{-\delta}$.

In the first case, because of (5),

$$f_N(\underline{a}(t)) = f_N(\underline{a} + t\,\underline{\Delta}_N)$$

(6) $$= f_N((1-t)\underline{a} + t\,\hat{\underline{a}}_N)$$

$$\leq f_N\,(\underline{a})$$

a consequence of (4), $t < 1$, and the convexity of f_N. Therefore,

(7) $P(N^{\delta}||\hat{\underline{a}}_N - \underline{a}|| > \epsilon) \leq P\{f_N(\underline{a}(t)) - f_N(\underline{a}) \leq 0\}$

Using (1) and (2), the event on the right in (7) has the same probability as

(8) $$E_N = \{(\sum_{i=1}^{N} (|U_{i+k} - V_i| - |U_{i+k}|)) \leq 0\}$$

where we write

$$(9) \qquad V_i = d_1X_{i+k-1} + \dots + d_kX_i.$$

From this observation, the next step is to write (8) in a form where it is easier to see that $P(E_N) \longrightarrow 0$. Using the fact that for $b > 0$

$$h_b(a) \equiv |a-b| - |a| = \begin{cases} b & \text{if } a < 0 \\ b-2a & \text{if } 0 \le a \le b \\ -b & \text{if } a > b \end{cases}$$

and $h_{-b}(a) = h_b(-a)$, we can write

$$|a-b| - |a| = b[\kappa(a < 0) - \kappa(a > 0)] +$$

$$2(b-a)[\kappa(0 \le a \le b) - \kappa(b \le a \le 0)],$$

$\kappa(A)$ denoting the indicator of A (<u>note</u>: $h_0(a) = 0$). Thus the summand in (8) may be written as

$$V_i[\kappa(U_{i+k} < 0) - \kappa(U_{i+k} > 0)] +$$

$$2(V_i - U_{i+k})\ [\kappa(0 \le U_{i+k} \le V_i) - \kappa(V_i \le U_{i+k} \le 0)]$$

and we have

(10) $$P(N^{\delta}||\underline{a}_N - \underline{a}|| > \epsilon) \le P(S_1 + S_2 \le 0),$$

where

(11) $$S_1 = \sum_{i=1}^{N} V_i[\kappa(U_{i+k} < 0) - \kappa(U_{i+k} < 0)]$$

and

(12) $$S_2 = 2\sum_{i=1}^{N} (V_i - U_{i+k})[\kappa(0 \le U_{i+k} \le V_i) - \kappa(V_i \le U_{i+k} \le 0)]$$

We will show that the right hand side of (10) converges to zero by showing $N^tS_1 \longrightarrow 0$ and $N^tS_2 \longrightarrow \infty$, both a.s., for an appropriate choice of t.

The argument for S_1 is quite easy. As $V_i = d_1X_{i+k-1} + ... + d_kX_i$, $S_1 = d_1T_1 + ... + d_kT_k$, where

(13) $$T_j = \sum_{i=1}^{N} Z_{i-j}$$

and $Z_{i-j} = X_{i+k-j}[\kappa(U_{i+k} \le 0) - \kappa(U_{i+k} \ge 0)]$, $j=1,...,k$. For any $\beta < \alpha$ and $i > 1$

$$
(14) \qquad
\begin{aligned}
(E(|Z_i|^{\beta}))^{1/\beta} &\le 2(E(|X_{i+k}|^{\beta})^{1/\beta}) \\
&\le 2\sum_{m=1}^{\infty}(E(|b_m\, U_{i+k-m}|^{\beta}))^{1/\beta} \\
&\le 2c\sum_{m=1}^{\infty} t^m (E(|U_{i+k-m}|^{\beta}))^{1/\beta} \\
&< \infty\,;
\end{aligned}
$$

the second line follows from $X_i = \sum b_j U_{i-j}$ and Minkowski's inequality; the third uses $|b_j| \le ct^j$, $t < 1$ and $E(|U_i|^{\beta}) < \infty$ for any $\beta < \alpha$, because U_i is attracted to a stable law.

In (13) the Z_{i-j} are already centered at conditional expectations. That is,

$$E(Z_{i-j}|Z_{i-j-1},\dots,Z_{i-j}) = 0$$

by the independence of U_{i+k} and X_{i+k-j}, $j > 0$ and the fact that

$$E(\kappa(U_{i+k} \le 0) - \kappa(U_{i+k} \ge 0)) = 0$$

because med $(U_i) = 0$. The result in (14) allows us to apply Marcinkiewicz's version of the strong law [Loeve (1963), p.242] which asserts

$$N^{-1/\beta}\, T_j \longrightarrow 0 \text{ a.s., } \beta < \alpha.$$

Thus, using $|d_i| \le \epsilon N^{-\sigma}$,

$$N^{-t} S_1 \longrightarrow 0 \text{ a.s., any } t > (1/\alpha) - \sigma \tag{15}$$

The argument for S_2 in (12) is more complex. Note that each term is non-negative, that eliminating $-\kappa(V_i \le U_{i+k} \le 0)$ reduces the sum, and that

$$\kappa(0 \le U_{i+k} \le V_i/2) \le \kappa(0 \le U_{i+k} \le V_i).$$

Hence

$$\begin{aligned} S_2 &\ge 2 \sum_{i=1}^{N} (V_i - U_{i+k})\ \kappa(0 \le U_{i+k} \le V_i/2) \\ &\ge \sum_{i=1}^{N} V_i\ \kappa(0 \le U_{i+k} \le V_i/2) \qquad (16) \\ &\ge \sum_{i=1}^{N} V_i\ \kappa(0 \le U_{i+k} \le V_i/2)\ \kappa(A_i) \end{aligned}$$

the last step trivially true for any event A_i. We will bound the summand on the last line of (16) from below by a quantity $Q_i > 0$ which still satisfies $N^t \sum Q_i \longrightarrow \infty$.

From Lemma 1.1, $X_m = \sum b_j U_{m-j}$, $b_0 = 1$. If this is used for each term in $V_i = d_1 X_{i+k-1} + \dots + d_k X_i$ we obtain, after simplification,

$$\begin{aligned} V_i &= (d_1) U_{i+k-1} + (d_1 b_1 + d_2) U_{i+k-2} + \dots \\ &\quad + (d_1 b_{k-1} + \dots + d_k) U_i + \sum_{j=1}^{\infty} (d_1 b_{k-1+j} + \dots + d_k b_j) U_{i-j} \end{aligned} \tag{17}$$

If we write

$$c_j = \sum_{i=1}^{\min(j,k)} d_i b_{j-i}, \quad j \geq 1,$$

(17) implies

$$\begin{aligned} V_i &= c_1 U_{i+k-1} + \dots + c_k U_i + \sum_{j=1}^{\infty} c_{k+j} U_{i-j} \\ &\geq -|c_1 U_{i+k-1}| \dots + c_j U_{i+k-j} - \dots - |c_k U_i| \\ &\quad - \sum_{j=1}^{\infty} |c_{k+j} U_{i-j}| \end{aligned} \tag{18$_j$}$$

for every $j = 1, \dots, k$. Since $\|\underline{d}\| = \epsilon N^{-\sigma}$ and $b_j < ct^j$ for some $t < 1$, $|c_{k+i}| \leq ck\epsilon N^{-\sigma} t^i$ for all i and $|c_i| \leq ck\epsilon N^{-\sigma}$, if $i \leq k$. This means that

$$V_i \geq c_j U_{i+k-j} - ck\epsilon N^{-\sigma}(\sum |U_{i+k-m}|)$$

$$(19)_j \qquad -ck\epsilon N^{-\sigma}(\sum_{m=1}^{\infty} t^m |U_{i-m}|)$$

the first sum over $m = 1, \ldots, k,\ m \neq j$.

The next step is to partition

$$D = \{\underline{d} \in R^k: ||\underline{d}|| = \epsilon N^{\delta}\}$$

so we can control both the sign and size of c_j on certain sets in the partition. Specifically, for $j = 1, \ldots, k$, write

$$(20) \qquad D_j = \left\{ \underline{d} \in D: \begin{array}{l} d_j > 3^{j-k-1} \epsilon N^{-\sigma} \text{ and} \\ |d_i| \leq 3^{i-k-1} \epsilon N^{-\sigma},\ i < j \end{array} \right\}$$

$$D_j^* = \left\{ \underline{d} \in D: \begin{array}{l} d_j < 3^{j-k-1} \epsilon N^{-\sigma} \text{ and} \\ d_i \geq 3^{i-k-1} \epsilon N^{-\sigma},\ i < j \end{array} \right\}$$

Then on D_j we have

$$c_j = d_j + d_{j-1}b_1 + \dots + d_1 b_{j-1}$$

$$\text{(21)} \qquad \geq d_j - \epsilon N^{-\sigma}(3^{j-2} + \dots + 1)/3^k \geq d_j/2$$

$$> 3^{j-1}\epsilon N^{-\sigma}/(2(3^k))$$

while on D_j^*,

$$c_j = d_j + d_{j-1}b_1 + \dots + d_1 b_{j-1}$$

$$\text{(22)} \qquad \leq d_j + \epsilon N^{-\sigma}(3^{j+2} + \dots + 1)/3^k \leq d_j/2$$

$$< 3^{j-1}\epsilon N^{-\sigma}/(2(3^k)).$$

Furthermore the D_j, D_j^* are mutually exclusive and cover D. If we knew that $\underline{d} \notin \bigcup_{i=1}^{k} (D_i \cup D_i^*)$

$$||\underline{d}|| \leq \frac{\epsilon N^{-\sigma}}{3^k}(9^{k-1} + \dots + 1)^{1/2} < \epsilon N^{-\sigma}/9$$

so $\underline{d} \notin D$.

Now given $K_1 > 0$ define for $j = 1,\dots,k$,

$$C_{ij}=\{N^{\sigma}<U_{i+k-j},\ \sum_{m=1}^{\infty} t^m|U_{i-m}|\le K_1,\ |U_{i+k-m}|\le K_1,\ m\le k,\ m\ne j\}$$

(23)

$$C^*_{ij}=\{-N^{\sigma}>U_{i+k-j},\ \sum_{m=1}^{\infty} t^m|U_{i-m}|\le K_1, |U_{i+k-m}|\le K_1,\ m\le k,\ m\ne j\}$$

events where U_{i+k-j} is big, U_{i+k-m} is bounded, $m \ne j$, and $\Sigma t^m|U_{i-m}|$ is bounded. From $(19)_j$, and (20),

$$V_i > (3^{j-k-1})\epsilon/2 - ck\epsilon(kK_3)$$

on $D_j \cap C_{ij}$ while on $D^*_j \cap C^*_{ij}$, the same bound holds, from (18). Thus, on $E_{ij} = (C_{ij} \cap D_j) \cup (C^*_{ij} \cap D^*_j)$ we finally obtain

(24) $$V_i > N^{-\sigma}(a_1 U_{i+k-j} - a_2),$$

as asserted, where $a_1 = \epsilon(2\cdot 3^k) > 0$ and $a_2 = ck^2\epsilon K_1 > 0$.

Returning to (16) we define

$$A_i = \bigcup_{j=1}^{k} E_{ij}.$$

Then

$$S_2 \geq \sum_{i=1}^{N} V_i \kappa(0 \leq U_{i+k} \leq V_i/2) \sum_{j=1}^{k} \kappa(E_{ij})$$

because of (16), and the fact that D_j, D_j^* partitions D. Reversing summation and using (24), we see that the right hand side is no more than

$$(25) \quad \sum_{j=1}^{k} N^{-\sigma} \sum_{i=1}^{N} (a_1 U_{i+k-j} - a_2) \kappa(E_{ij}) \; \kappa(0 \leq U_{i+k} \leq N^{-\sigma}(a_1 U_{i+k-j} - a_2))$$

By (23) $|U_{i+k-j}| > N^{\sigma}$ on E_{ij} and, putting it all together, we find that the pth term in the outer sum (over j) in (25) is

$$(26) \quad \begin{aligned} W_p &\geq K_2 \sum_{i=1}^{N} \kappa(E_{ip}) \kappa\,(0 \leq U_{i+k} \leq K_2) \\ &= K_2 \sum_{i=1}^{N} [\kappa(C_{ip})\kappa(D_p) + \kappa(C_{ip}^*)\kappa(D_p^*)] \; \kappa(0 \leq U_{i+k} \leq K_2) \\ &= K_2 \kappa(D_p) \sum_{i=1}^{N} \kappa(C_{ip}) \kappa(0 \leq U_{i+k} \leq K_2) \\ &+ K_2 \kappa(D_p^*) \sum_{i=1}^{N} \kappa(C_{ip}^*) \kappa(0 \leq U_{i+k} \leq K_2) \end{aligned}$$

for some $K_2 > 0$.

From (23) C_{ip} is the intersection of three events,

$$\{U_{i+k-p} > N^{\sigma}\},\ \{\sum_{m=1}^{\infty} t^m |U_{i-m}| \le K_1\},\ \cup\{|U_{i+k-m}| \le K_1\},$$

the last union over $m=1,\dots,k$, $m \ne p$. They are independent because the U_i's are and the subscripts don't overlap. Also they are independent of $\{0 \le U_{i+k} \le K_2\}$. Similarly for C^*_{ip}. Thus, by the ergodic theorem, if N is large enough

$$N^{-1}W_p + \gamma \ge K_2 \kappa(D_p) E(C_{ip})\ P\{U_{i+k} \in [0,K_2]\}$$
$$+ K_2 \kappa(D^*_p) E(C^*_{ip})\ P\{U_{i+k} \in [0,K_2]\}$$

for any $\gamma > 0$. Let $K_3 = P\{U_{i+k} \in [0,K_2]\}$. It is positive because $0 = \text{med}(U_i)$. Note also that $E(C_{ip}) = K_4 P\{U_{i+k-p} > N^{\sigma}\}$, where K_4 is the probability of the event

$$\{\sum_{m=1}^{\infty} t^m |U_{i-m}| \le K_1 \text{ and } |U_{i+k-m}| \le K_1,\ m=1,\dots,k,\ m \ne p\};$$

clearly $K_4 > 0$.

Since the U_i are attracted to a stable law of index $a > 1$, $E(C_{ip}) \sim K_5\ N^{-\sigma a}$ so $N^{-1+a\sigma+\delta}\ W_p \longrightarrow \infty$ a.s. for all $\delta > 0$. This shows that

(27) $N^{-1+a\sigma+\delta}\ S_2 \longrightarrow \infty$ a.s., $\delta > 0$

As $a > 1$ and $\sigma < 1/a$, $a(1-a\sigma) > 1-a\sigma$ and $1-a\sigma > 1/a-\sigma$. If we put $t=[(1-a\sigma)+(1/a-\sigma)]/2$ and $\delta = 1-a\sigma-t$, then both $t > 1/a-\sigma$ and $\delta > 0$. (15) and (27) imply $N^{-t}S_1 \longrightarrow 0$ a.s. and $N^{-t}S_2 \longrightarrow \infty$ a.s., which, together with (10), completes the proof.■

There is the possibility that a more rapid rate of convergence holds for LAD autoregression estimates. In fact it is even suggested. The Monte-Carlo experiments described in the next section imply a different, faster convergence rate for LAD than for least squares.

Another possible modification of Theorem 2 is to extend the range of applicability of the stated rate. For example, it is not difficult to establish

Theorem 3: If the U_i are Cauchy distributed and centered at zero then for all $\delta < 1$ ($= 1/a$),

(28) $N^{\delta}(\hat{\underline{a}}_N - \underline{c}) \longrightarrow 0$ in probability

Proof: The argument that $N^{-t}S_2 \longrightarrow \infty$ a.s. for $t < 1-\sigma$ still goes through as in the proof of Theorem 2, since none of the limit theorems invoked $\alpha > 1$. We need only consider S_1 in (11).

The trick is to use $X_i = \sum b_j U_{i-j}$ in

$$(29) \qquad N^{-1} \sum_{i=1}^{N} X_{i+k-j}[\kappa(U_{i+k}>0)-\kappa(U_{i+k}<0)].$$

Since the X's are stationary Cauchys, and the terms in (29) are mixing, the limit law is Cauchy [see Ibragimov and Linnik (1971) for the analogue of the central limit theorem for stable laws]. It follows that

$$N^{-t}|S_1| \leq kN^{-t} N^{-\sigma} \left| \sum_{i=1}^{N} X_{i+k-j}[\kappa(U_{i+k}<0) - \kappa(U>0)] \right|$$
$$= k\, N^{-(t+\sigma-1)}C,$$

C a Cauchy random variable, and the right hand side $\to 0$ in probability if $t > 1-\sigma$. The final choice of t is made just as it was done in the proof of Theorem 2, and this completes the proof of Theorem 3.∎

The proof of Theorem 2 depends on $\alpha > 1$ for the ergodic theorem. The statement may remain true even if $\alpha < 1$, but a different technique would probably be required to demonstrate it. The experiment of the next section actually suggests rather strongly that this may indeed be the case.

3.3 Sampling Behavior of LAD

In this section we describe a Monte-Carlo experiment which illustrates the behavior of LAD autoregression estimates. This behavior is compared to that of least squares and other Huber M-estimators.

We study the model

$$(1) \qquad X_{n+1} = 1.4\ X_n - .7\ X_{n-1} + U_{n+1}$$

which is stationary because $1.4z - .7z^2 - 1$, has roots $1 \pm ((1.4)^2 - 2.8)^{1/2}/1.4$, both outside the complex unit circle $|z| \leq 1$. Thus by Lemma 1.1, there exists an a.s. unique stationary process that satisfies (1).

We used 6 different distributions for U_i. Two of them were finite variance distributions, the first being standard normal, the second, double exponential.

Two distributions are in Dom (1.2). The first, as in (2.5.2), is

$$(2) \qquad h_{.05}(t) = \begin{cases} (.95e^{-t^2/2})/(2\pi), & |t| < k \\ c/(2(1+|t|)^{2.2}), & |t| \geq k \end{cases}$$

which modifies the normal by $P_{1.2}$ (see (2.5.3)). Here k(.05) = 1.399 and c is chosen so f has integral 1 (approximately 2.4). This makes U_i normal, given that $U_i \in [-k(.05), k(.05)]$, and Pareto $P_{1.2}$ outside. The density $h_{.05}$ is differentiable except at ± 1.399 and has a unique median at 0, so U_i satisfies the requirements of Theorem 2.1 and Theorem 2.2.

The other Dom (1.2) density is $g_{1.2}$ from (2.5.3). It is everywhere differentiable except at 0, its unique median. Again it makes U_i satisfy the assumptions of the Theorems of Section 2.

Finally, two distributions attracted to a stable law of index $\alpha = .6$ were used in the experiment. Both have infinite mean. If U_i in (1) follows one of these densities, the behavior of the LAD estimator for (1) need not obey the foregoing theorems. The first such density is

$$(3) \qquad s_{.05}(t) = \begin{cases} (.95e^{-t^2/2})/(2\pi), & |t| < k \\ c/(2(1+|t|)^{1.6}), & |t| \geq k \end{cases}$$

where again, $k(.05) = 1.399$, and c is chosen so $s_{.05}$ has integral 1, (approximately 1.7). As in the case of (2), $s_{.05}$ is differentiable everywhere except at $\pm$ 1.399 and has unique median 0. The second infinite mean distribution is $g_{.6}$ from (2.5.3.), differentiable except at its median, 0.

In summary then, the six autoregressions we studied vary according to which distribution of errors, U_i, was employed:

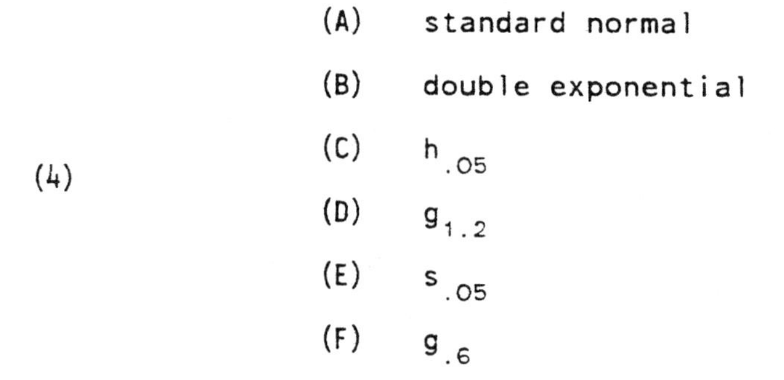

(4)

(A) standard normal
(B) double exponential
(C) $h_{.05}$
(D) $g_{1.2}$
(E) $s_{.05}$
(F) $g_{.6}$

The error distributions in the present study explore the finite variance case much less than the Chapter 2 experiments did. This is intentional. For one thing, the error distribution of U_{n+1} affects the distribution of the design variables

$X_n, X_{n-1}, \ldots, X_{n-k}$, in (1) making it is difficult to isolate these effects. Secondly, the behavior of LAD, and the way it may contrast with that of other estimators is likely to be more sensitive to the tails of the U_i than to aspects of the distribution shape.

For each error distribution we obtained N+2 observations of the process in (1) as follows: we generated

$$U_{-300}, U_{-299}, \ldots, U_0, \ldots, U_{N+2},$$

independently from a distribution in (4). Then we approximated X_i by

$$X_i^* = \sum_{j=0}^{i+300} b_j U_{i-j}, \tag{5}$$

the b_i obtained from $a_1 = 1.4$, $a_2 = -.7$ as in (1.2). Also from (1.2),

$$X_i - X_i^* = \sum_{j=1}^{k} c_j X_{-300-j}.$$

Because $|c_j| < K(t)^{300}$ is extremely small, X_i^* and X_i are so close that X^* must satisfy (1) up to the first several decimal places.

Using $X_1^*,\ldots,X_{n+2}^*$ the least squares estimator (1.7), the LAD estimator and two Huber M–estimators were computed. The latter finds $\hat{\underline{c}}$ to minimize

$$(6) \qquad H_\delta(\underline{c}) = \sum_{i=1}^{N} \rho(X_{i+k} - (c_1 X_{i+k-1} + \ldots + c_k X_i))$$

where ρ is defined by

$$(7) \qquad \rho(t) = \begin{cases} t^2/2, & |t| \le \delta \\ \delta|t| - \delta^2/2, & |t| > \delta \end{cases}$$

This context is similar to that of Section 2.5. If $\delta > 0$ is large enough, ρ weights all non–zero residuals quadratically, so the minimizer of H_δ is least–squares. This is maximum likelihood for location when the errors are contaminated by an amount $\epsilon=0=k^{-1}(\infty)$ and, following the terminology introduced in Section 2.5, we call the least squares estimators $\hat{\underline{c}}_0$.

Similarly if $\delta > 0$ is small enough, H_δ treats each non–zero residual linearly, and is LAD. If we were estimating location of variables with densities f_ϵ, LAD would be maximum likelihood under the contamination rate $\epsilon = 1 = k^{-1}(0)$. Again, as in Section 2.5, we suggestively denote LAD as $\hat{\underline{c}}_1$.

Finally, though the notation does not appeal to the same extent it did in Section 2.5, we call the minimizer of $H_{1.399}$, $\hat{c}_{.05}$ and that of $H_{.766}$, $\hat{\underline{c}}_{.25}$. Clearly the ordering

$$\underline{c}_0 << \underline{c}_{.05} << \underline{c}_{.25} << \underline{c}_1$$

suggests the relative degree of protection against the effects of large residuals that is offered by these four Huber M-estimators.

Three sample sizes, N, were utilized, N = 50 for small samples, N = 150 for moderate samples, and N = 300 for larger samples. In the first two cases, 301 replications were performed. In each, $X_1^*,...,X_{n+k}^*$ was generated and then $\hat{\underline{c}}_0$, $\hat{\underline{c}}_{.05}$, $\hat{\underline{c}}_{.25}$, $\hat{\underline{c}}_1$, computed and the errors of estimation were recorded. For N = 300 only 101 replications were done.

A summary of the results may be seen in the following table. For $\underline{a}=(1.4,-.7)$ it gives the median length, $||\hat{\underline{c}}-\underline{a}||$ over the 301 replications (101 when N=300) for each of the 4 estimators, for each of the six sampling situations listed in (4)

Table 1

Median length of the error vector $\hat{\underline{c}}-\underline{a}$ for various estimates of $\underline{a}$ in (1), under various error distributions, and various sample sizes N.

N=50(301 reps.)

	A	B	C	D	E	F
$\hat{\underline{c}}_0$	.1077	.0981	.0901	.0604	.0374	.0232
$\hat{\underline{c}}_{.05}$	.1167	.0887	.0587	.0212	.0112	0019
$\hat{\underline{c}}_{.25}$	.1243	.0877	.0538	.0176	.0104	.0017
$\hat{\underline{c}}_1$	.1429	.0820	.0608	.0155	.0104	0014

N=150(301 reps.)

	A	B	C	D	E	F
$\hat{\underline{c}}_0$	.0596	.0605	.0395	.0364	.0104	.01076
$\hat{\underline{c}}_{.05}$	.0617	.0494	.0248	.0093	.0018	.00037
$\hat{\underline{c}}_{.25}$	.0666	.0435	.0217	.0081	.0015	.00028
$\hat{\underline{c}}_1$	.0733	.0433	.0232	.0063	.0017	.00027

N=300(101 reps.)

	A	B	C	D	E	F
$\hat{\underline{c}}_0$	.0361	.0378	.0334	.0227	.00438	.00333
$\hat{\underline{c}}_{.05}$	.0397	.0307	.0161	.0043	.00045	.00010
$\hat{\underline{c}}_{.25}$	.0438	.0311	.0166	.0036	.00044	.00010
$\hat{\underline{c}}_1$	.0512	.0312	.0181	.0032	.00042	.00009

Several regularities are manifest. First, errors decrease in each row, moving from left to right, except that for N=150, E<<F.

For normal autoregressions, the ordering

$$\underline{c}_0 << \underline{c}_{.05} << \underline{c}_{.25} << \underline{c}_1$$

is seen for each sample size, as expected. Similarly, the reverse ordering of estimators holds when errors are double exponential.

When the errors are in Dom (1.2) (cases C and D, with finite mean but infinite variance) least squares estimators produce the largest average errors. In the pure Pareto $P_{1.2}$ case (D), the ordering $\underline{c}_1 << \underline{c}_{.25} << \underline{c}_{.05} << \underline{c}_0$ is shown, as expected. Similarly with the $P_{.6}$ case in column F.

Now, focusing on a particular error distribution, and comparing average error sizes as N increases, the rate of convergence to zero is suggested. In particular in cases C–F it seems that $||\hat{\underline{c}}_0 - \underline{a}||$ goes to zero slower than $||\hat{\underline{c}}_1 - \underline{a}||$, so LAD may be asymptotically more efficient then least squares. In fact, having compared the ratio of $||\hat{\underline{c}}_1 - \underline{a}||$ to $||\hat{\underline{c}}_0 - \underline{a}||$ over 301 replications at each of the sample sizes we conjecture

$$||\hat{\underline{c}}_1 - \underline{a}|| / ||\hat{\underline{c}}_0 - \underline{a}|| \rightarrow 0, \tag{8}$$

and at a rate that increases as α decreases. This assertion is supported by Table 2.

Some details of the Monte-Carlo results suggest that (8) may also hold for the other Huber M-estimators, $\hat{\underline{c}}_{.05}$, $\hat{\underline{c}}_{.25}$. It is not clear how to go about studying these statements.

Table 2

Median of 301 values of
$$||\hat{\underline{c}}_1-\underline{a}||/||\hat{\underline{c}}_0-\underline{a}||$$
for various error distributions and sample sizes.

	$\alpha=2$		$\alpha=1.2$		$\alpha=.6$	
	A	B	C	D	E	F
N=50	1.214	.817	.767	.261	.367	.073
N=150	1.172	.733	.653	.212	.207	.028
N=300	1.356	.764	.615	.143	.153	.031

3.4 Notes

1. In the classical case of (1.1), $E(U_i^2) < \infty$ is assumed [this certainly implies (1.5)]. With stationarity, (3), $\sum b_j U_{n-j}$ converges in L_2, therefore in probability. Kanter and Steiger(1974) gave the first generalization to the case where U_i is attracted to a non-normal stable law of index α. The condition (1.5) and Lemma 1.1 is from Yohai and Maronna (1977).

Being in dom(α) is much more restrictive than the Yohai-Maronna condition, (1.5). However they impose, though don't use, symmetry of U_i.

2. The theorem in (1.9) began in Kanter and Steiger (1974) with $\hat{\underline{c}} \longrightarrow \underline{a}$ in probability. At the 1974 Brasov Conference [Kanter and Steiger (1977)] they noticed that their consistency proof actually implied the rate $N^\delta(\hat{\underline{c}}_N - \underline{a}) \longrightarrow 0$ in probability, $\delta < \min(1/\alpha, (2-\alpha)/\alpha)$. Kanter and Hannan (1977) cleaned up the bound on δ and established convergence with probability 1. Under a weaker condition than $U_i \in \mathrm{Dom}(\alpha)$, namely (1.5), Yohai and Maronna established a weaker result, namely that $N^\delta(\hat{\underline{c}}_N - a) \longrightarrow 0$ in probability, $\delta < 1/2$. Thus a wider class of processes is embraced at the expense of weakening the convergence rate. Again, we cite Feller (1971) as a convenient source for details about stable laws.

3. Theorem 2.1 is from Gross and Steiger (1979). The statement of Theorem 2.2 was one of two conjectures they made.

4. Theorems 2.2 and 2.3 are from An and Chen (1982). The proof that $N^{-t}S_2 \rightarrow \infty$ in (1.12) is similar to theirs. Our proof that $N^{-t}S_1 \rightarrow 0$ in (1.11) is much simpler. Curiously, they give an example, which, if correct, would seem to invalidate their method of proof of Theorem 2.2.

5. All that is needed for Theorem 2.3 is that N_p times the expression in (2.29) converge to zero. The original proof of this is more complicated than ours.

4. LAD IN MULTI-WAY TABLES

4.1 One-way layout

An important special case of the general linear model discussed in Chapters 1 and 2 is when the data fall into a multi-way table. The simplest case is the one-way layout, where the data are organized into c cells, with observations y_{jk}, $1 \leq k \leq n_j$, in the jth cell, $1 \leq j \leq c$.

If the cells are to be analyzed individually, the one-way structure adds no complications. However, we often wish to decompose such data according to

$$(1) \qquad y_{jk} = \mu + \beta_j + r_{jk} \qquad \begin{array}{l} 1 \leq k \leq n_j \\ 1 \leq j \leq c \end{array} ,$$

where μ represents the general overall level of the observations, and β_j represents the general deviation of observations in cell j from the overall level. The parameters $\{\beta_j : 1 \leq j \leq c\}$ are also called <u>effects</u> associated with the cells [Tukey (1977)].

One way to estimate the parameters μ and β_j, $1 \leq j \leq c$ is by the LAD criterion: minimize

$$\sum_{j=1}^{c} \sum_{k=1}^{n_j} |y_{jk} - \mu - \beta_j|.$$

This is easily done by fixing μ to be any convenient value $\hat{\mu}$, and then choosing β_j to minimize the jth inner sum. Clearly $\hat{\mu} + \hat{\beta}_j$ is a median of $\{y_{j,k}: 1 \leq k \leq n_j\}$, and may be non-unique if n_j is even. More serious nonuniqueness arises because $\hat{\mu}$ was chosen arbitrarily.

The latter form of nonuniqueness arises also in the least squares analysis of such data and is usually eliminated by adding the constraint $\sum n_j\beta_j = 0$ (Johnson and Leone, 1964, p. 6). In the context of LAD methods, this would amount to estimating μ by the weighted average of cell medians with weights proportional to n_j. However, to preserve the insensitivity of the analysis to disturbances of the data, it seems more appropriate to use the weighted median.

We note that requiring $\sum n_j\hat{\beta}_j = 0$ is equivalent to minimizing $\sum n_j\hat{\beta}_j^2$, subject to $\hat{\mu} + \hat{\beta}_j$ being held fixed. Similarly, taking $\hat{\mu}$ to be the weighted median is equivalent to solving the problem

$$\text{minimize} \quad \sum_{j=1}^{c} n_j |\beta_j|$$

$$\text{subject to} \quad \sum_{j=1}^{c} \sum_{k=1}^{n_j} |y_{jk} - \mu - \beta_j| = \text{minimum}$$

Armstrong and Frome (1979) discuss a similar but unweighted solution to the problem. Either approach will eliminate the complete arbitrariness in the choice of μ but may leave an interval of possible values. Finally Armstrong and Frome suggest choosing $\hat{\mu}$ to be the value closest to the median of all the observations and claim that this makes the situation unique. However, this is only true if we agree to adopt a unique definition of the median of the entire set of observation, in the case that $n = \sum n_j$ is even! The conventional choice is the midmedian, namely the average of the two middle order statistics, which seems reasonable. However, we point out that nonuniqueness has only been finally eliminated by introducing some criterion beyond that of LAD.

The Armstrong-Frome solution has the curious property that $\hat{\mu}$ is equivariant under the addition of a constant to all observations, but that $\hat{\mu} + \hat{\beta}_j$ is not necessarily equivariant under the addition of a constant to observations only in cell j. If n_j is even and the median of cell j is nonunique, then $\hat{\beta}_j = 0$ for any translation of the cell for which the median interval contains $\hat{\mu}$. Full equivariance could be restored by requiring

$\hat{\mu} + \hat{\beta}_j$ to be the unique median of cell j. However, in multi-way layouts we shall encounter nonuniqueness of a more subtle form, arising from other aspects of the structure of the data than parity of cell size. We believe it is better to eliminate nonuniqueness by hierarchically

$$\text{minimizing } \sum \sum |y_{jk} - \mu - \beta_j|$$

then, subject to this,

$$\text{minimizing } \sum n_j |\beta_j|$$

then, subject to this

$$\text{minimizing } |\mu|.$$

As Armstrong and Frome remark, such hierarchical optimizations can be carried out by adding the three criteria with successively smaller appropriately chosen weights.

To see that this sequence of minimizations necessarily gives a unique solution, it is enough to note that the first stage requires each $\mu + \beta_j$ to lie in some closed interval, and that the second stage therefore requires μ to lie in a closed interval. The final stage therefore selects one end-point of the latter interval, or zero.

We remark that rules such as these may give unexpected results when many cells contain a small, even number of observations. For example if $n_j = 2$, $j = 1,\dots,c$, we only require that $\mu + \beta_j$ lies between the two values in the cell, and the particular value will be determined by the latter stages of optimization.

As an example, consider the following set of data [Table 4.4, Fisher and McDonald, (1978)]. Here $c = 3$, $n_1 = 2$, $n_2 = 4$, $n_3 = 4$, and

$$y_{j,k} = \begin{cases} 7.3,\ 7.4 & j = 1,\ k = 1,2 \\ 13.3,\ 10.6,\ 15.0,\ 20.7 & j = 2,\ k = 1,\dots,4 \\ 14.7,\ 23.0,\ 22.7,\ 26.6 & j = 3,\ k = 1,\dots,4 \end{cases}$$

We first find

$$7.3 \le \hat{\mu} + \hat{\beta}_1 \le 7.4$$

$$13.3 \le \hat{\mu} + \hat{\beta}_2 \le 15.0$$

$$22.7 \le \hat{\mu} + \hat{\beta}_3 \le 23.0$$

Minimizing $2|\hat{\beta}_1| + 4|\hat{\beta}_2| + 4|\hat{\beta}_3|$ subject to these inequalities, we find $\hat{\mu} = 15.0$, and $\hat{\beta}_1 = -7.6$, $\hat{\beta}_2 = 0$, $\hat{\beta}_3 = 7.7$. If we form a reduced data set by dropping the two extreme values from each of cells 2 and 3, we find the same inequalities on $\hat{\mu} + \hat{\beta}_j$, $j = 1, 2, 3$, but we must now minimize $2|\hat{\beta}_1| + 2|\hat{\beta}_2| + 2|\hat{\beta}_3|$, which leaves us with the solution

$$13.3 \le \hat{\mu} \le 15.0,$$

$$\hat{\beta}_1 = 7.4 - \hat{\mu}$$

$$\hat{\beta}_2 = 0$$

$$\hat{\beta}_3 = 22.7 - \hat{\mu}.$$

Finally minimizing $|\hat{\mu}|$, we obtain $\hat{\mu} = 13.3$, $\hat{\beta}_1 = -5.9$, $\hat{\beta}_2 = 0$, and $\hat{\beta}_3 = 9.4$. Notice that $\hat{\mu}$ moves from one end to the other of the median interval for cell 2, depending on the relative weights of cells 1 and 3.

4.2 Two-way layout

We now consider data recorded in cells with a higher level of classification structure, which we exemplify with the two-way layout. Our data may now be described by

$$(1) \qquad y_{i,j,k}: \begin{cases} 1 \le k \le n_{ij} \\ 1 \le i \le r \ , \ 1 \le j \le c \end{cases}$$

where r and c are the numbers of rows and columns in the table, respectively, and n_{ij} is the number of observations in the (i,j) cell.

The closest analogue to the decomposition (1.1) is the representation

$$(2) \qquad y_{i,j,k} = \mu + \alpha_i + \beta_j + \gamma_{ij} + r_{ijk}$$

Since this allows the typical value $\mu + \alpha_i + \beta_j + \gamma_{ij}$ in the (i,j) cell to vary freely, the estimation by LAD again amounts to requiring this value to lie in the median interval of observations in the (i,j) cell (which may reduce to a single point). Thus the only issue is how to resolve the resulting nonuniqueness. Again, we suggest progressively minimizing

$$\sum_{i,j} n_{ij} |\gamma_{ij}|,$$

then

$$\sum_{i,j} n_{ij}(|\alpha_i| + |\beta_j|)$$

then

$$|\mu| ,$$

always requiring previous criteria to remain at their minima.

Another decomposition of two-way data that is often of interest is the interaction-free or "row-plus-column" decomposition

$$(3) \qquad y_{ijk} = \mu + \alpha_i + \beta_j + r_{ijk}$$

Since the typical value for cell (i,j) is no longer free to vary independently of all other cells, the estimation procedure is not as straightforward. However, (3) is easily put in the form of the general linear model discussed in Chapters 1 and 2, and therefore estimates may be found using the algorithms described in Chapters 1 and 7. These algorithms only find <u>one</u>

solution and would need minor modification to find the particular solution that satisfies the additional criteria to guarantee uniqueness, namely

$$\text{minimize } \sum n_{ij}(|\alpha_i| + |\beta_j|)$$

and then

$$\text{minimize } |\mu| .$$

The simplest, though not necessarily the most efficient way to find this specific solution is effectively to add r+c+1 pseudo-observations with the value zero, in pseudo-cells for which the expected values are $\epsilon\sum_j n_{ij}\alpha_i$, $1 \le i \le r$, $\epsilon\sum_i n_{ij}\beta_j$, $1 \le j \le c$, and $\epsilon^2\mu$, respectively, for a sufficiently small positive ϵ.

4.3 Properties of LAD estimates

The general properties of LAD estimates that were established in Chapters 1 and 2 all apply in this context. For instance, Theorem 1.2.1 assures us that in analyzing any table by LAD, there exists a solution with at least as many zero residuals as there are degrees of freedom in the fit. These are c, rc and r+c−1, respectively, in the cases of decompositions (1.1), (2.2) and (2.3). The first two results are obvious since c and rc are the numbers of cells in one−way and two−way tables, respectively. The third value is not obvious but follows from dimensionality arguments. [See Siegel (1983), for example].

To apply the statistical results of Chapter 2, we have to make some assumptions about the probability mechanism generating the data. These are:

i) that the observations in each cell satisfy the decomposition (1.1), (2.2) or (2.3), whichever is appropriate, where the residual r is a random variable U with unique median at zero;

ii) that the number of observations in the cells are multinomially distributed, in such a way that property (2.2.2)is satisfied.

In the case of the one−way layout or the two−way layout with interactions, property (2.2.2) is satisfied if and only if there is a positive probability associated with each cell. In the case of (3), the two−way layout with no interactions, we only need a positive probability on r+c−1 cells whose fits are linearly independent.

With these assumptions, Theorems 2.2.1 and 2.2.2 assure us of almost sure convergence and asymptotic normality of the LAD estimates, respectively. (Strictly speaking, this is assured only if we use linear constraints to eliminate nonuniqueness.)

4.4 LAD and Median Polish

In the "row-plus-column" model

$$y_{ijk} = \mu + \alpha_i + \beta_j + r_{ijk},$$

the LAD criterion seeks a minimizer of the norm of the table of residuals

$$(1) \qquad \sum_{i,j,k} |y_{ijk} - \mu - \alpha_i - \beta_j|.$$

It is easily seen that if the median of the residuals in all cells in any row or column is not zero, then the fit does not minimize (1). The corresponding α or β can be adjusted to make the median residual zero, thereby reducing one part of the sum (1) without changing the value of the complementary part.

Therefore, the LAD fit has the property that the residuals have zero median in each row and in each column.

Thus there is a relationship between LAD fitting and the technique known as median polish [Tukey (1977)]. This is a tool for exploratory analysis of two-way tables. It has previously been described only for tables with no replication (n_{ij} = 1, all i,j), but there are no difficulties in extending it to the case of replication. One "half-step" of the process consists of subtracting row (or column) medians from the entries in the table, at the same time adding the same values to a set of effects. Alternate "half-steps" operate on rows and on columns, and the process is repeated until it converges.

For instance, consider the 3 x 3 table

(2)

1	2	4
3	3	5
4	5	5

First we border the table by zeroes, the initial values of the effects, α_i, β_j, N:

1	2	4	0
3	3	5	0
4	5	5	0
0	0	0	0

This has norm 24. To median polish the table beginning with rows (columns may also be used), we operate row by row, subtracting the row median from each entry in the row in the body of the table, and *adding* the median to the bordering entry. The result is

-1	0	2	2
0	0	2	3
-1	0	0	5
0	0	0	0

The second half-step operates on columns, *including* the bordering column, and results in

0	0	0	-1
1	0	0	0
0	0	-2	2
-1	0	2	3

with norm 3. The row medians were not changed from zero by this half-step, and consequently the process has converged: the table is *polished*. More usually, several steps are required before all row and column medians are zero, which is the condition for a polished table, and for convergence of the method. Median polish is described in greater detail by Emerson and Hoaglin (1983).

At each stage, the tableau gives a decomposition of the data into *effects* and *residuals*. For instance, the entry 5 in the lower right hand corner of the original table in (2) is at each stage the sum of the corresponding residual, the third row effect, the third column effect and the overall common term in the lowest right hand corner. For this set of data, the final effects and residuals of the median polish turn out to be the same as those for the LAD fit of the model (2.3).

By definition, at convergence the "polished" table of residuals necessarily shares with LAD the property that every row and column of residuals has zero median. However, this property does not imply that the result of median polish is a solution to the LAD problem in (1). Siegel (1983) gives the example of the table

$$\begin{array}{rrr} 1 & 6 & 3 \\ 5 & 9 & 2 \\ 6 & 4 & 7 \end{array} \tag{3}$$

for which median polish starting with columns converges to a set of residuals and effects given by

$$\begin{array}{rrr|r} -4 & 0 & 0 & 0 \\ 0 & 3 & -1 & 0 \\ 0 & -3 & 3 & 1 \\ \hline 0 & 1 & -2 & 5 \end{array}$$

with norm 14, while starting with rows the polished table is

$$\begin{array}{rrr|r} -2 & 0 & 0 & -3 \\ 0 & 1 & -3 & 0 \\ 0 & -5 & 1 & 1 \\ \hline -1 & 3 & 0 & 6 \end{array}$$

with norm 12. A LAD solution for this table, however, has residuals and effects

$$\begin{array}{rrr|r} -1 & 0 & 0 & -3 \\ 0 & 0 & -4 & 0 \\ 0 & -6 & 0 & 1 \\ \hline -1 & 3 & 0 & 6 \end{array}$$

with norm 11. Thus we cannot regard median polish as an algorithm for obtaining LAD fits.

However, Siegel gives a result which indicates that a <u>modified</u> median polish <u>is</u> guaranteed to converge to a LAD solution in the case of 3 x 3 tables. The details are in

> **Theorem 1:** If the first half step of median polish in a 3 x 3 table results in a row or column of zero residuals, then median polish will converge in at most 3 steps to a LAD solution.

Suppose however that we have a table like the example just given, where starting neither by rows nor by columns do we obtain a complete row or column of zeroes. Theorem 1 can still apply if we observe that by adding a constant to all the entries of a given column (say), we perturb the problem in a very transparent way. Namely, the corresponding β is changed by the same constant, and the residuals are unchanged. Thus we could modify the table in (3) to

101	6	203
105	9	202
106	4	207

and guarantee that the median of each row occurs in the first column. Median polish starting with rows now converges in 3 half steps to the solution

-1	0	0	-3
0	0	-4	0
0	-6	0	1
0	-96	101	105

This corresponds in an obvious way to the LAD solution given above for the table in (3).

Theorem 1 applied because this modification forces the algorithm to make all entries in one column zero at the first half-step. However, it would be more direct to modify the algorithm itself to achieve this, instead of manipulating the algorithm by modifying the data. We shall call "modified median polish" the algorithm obtained by specifying that in the first half-step the value subtracted from each row (or column) shall be _first_ entry in that row (or column), instead of the _median_ entry. We then have immediately

> **Theorem 2:** For any 3 x 3 table, modified median polish will converge in at most 3 steps to an LAD solution.

The performance of modified median polish has not been studied in larger tables, however, and we therefore cannot regard it as an algorithm for obtaining LAD fits except in the 3 x 3 case.

Siegel (1983) also shows that median polish of a table with rational entries always converges in a finite number of steps, provided the median of an even number of values is taken to be the low median. The low median is defined to be the lower of the two middle order statistics. It is easy to construct examples where the more conventional definition, the midpoint of the median interval, leads to geometric convergence, not convergence in a finite number of steps.

However, even for "low median" analysis of tables with rational entries, the number of iterations required can be arbitrarily large. Siegel gives the example of the 5 x 5 table

$$\begin{matrix} 4+\epsilon & 3 & 2 & 2 & 1 \\ 5 & 4 & 3 & 3 & 2 \\ 6 & 5 & 4 & 3 & 2 \\ 6 & 5 & 5 & 4 & 3 \\ 7 & 6 & 6 & 5 & 4 \end{matrix}$$

where ϵ is a small positive rational number. He shows that median polish starting with rows takes around $4/\epsilon$ steps to converge to a solution in which the residuals are

$$\begin{matrix} \epsilon & 0 & -1 & 0 & 0 \\ 0 & 0 & -1 & 0 & 0 \\ 1 & 1 & 0 & 0 & 0 \\ 0 & 0 & 0 & 0 & 0 \\ 0 & 0 & 0 & 0 & 0 \end{matrix}$$

(Note that since both r and c are odd, median polish and "low median" polish are the same). There is a sharp contrast here with most LAD algorithms, for which an upper bound on the number of iterations can be found from the number of observations and the dimension of the space of fits, independently of the values of the observations.

Kemperman (1983) has also discussed the relationship between LAD and median polish. He uses the term EMP (for end product of median polish) to describe a table of residuals whose row and column medians are all zero, and EMMP for the case where the midmedians are also zero. Also he calls a *weak* EMMP any table for which the pattern of signs is the same as that of an EMMP. Thus a row (or column) with an even number of entries must either have a unique median value of zero, or else exactly half the entries must be positive and half negative. However, in the latter case the midmedian is *not* required to be zero. Finally he calls a pair (r,c) "safe" for median polish (or midmedian polish) if every EMP (or EMMP) of dimension r x c is optimal in the LAD sense. The main result linking these ideas is

> **Theorem 3** (Theorem 6 of Kemperman (1983) No pair (r,c) with $r \geq 2$, $c \geq 2$ is safe for median polish. The only pairs which are safe for mid-median polish are the exceptional pairs: (2,c); (3,4); (4,4); (4,5); (4,6) and their reflections such as (r,2).

> In fact, for these exceptional pairs it is even true that every weak EMMP of dimension r x c is optimal.

Thus the problems that we have illustrated with 3 x 3 tables are also seen in tables of higher dimension.

To summarize, we have shown that there is a close connection between median polish and LAD fitting to tables. However, the lack of a guarantee of obtaining an LAD fit means that median polish could only be used to find a good starting point for a LAD algorithm, and the unboundedness of the number of iterations means that it could take a long time to get there.

We believe that median polish is very valuable as a quick, often hand-calculated way of summarizing data organized into a table. But we do not see it playing a central role in obtaining LAD fits for tables.

4.5 Nonuniqueness of "row-plus-column" fits

We have pointed out that the LAD "row-plus-column" fit to a table is often nonunique. The similarity to median polish makes this quite plausible when there is an even number of observations in any row or column. However, the nonuniqueness in the case of two-way tables comes from a more subtle structural property and is not tied to parity. For instance, the 2x2 table with multiple cell entries

1,2	3
4	5,6

may be decomposed as

0,1	0	0
0	-1,0	3
0	2	1

or as

-1,0	0	0
0	0,1	2
0	1	2

both of which are LAD fits. Note that both rows and both columns contain three data points.

Neither is nonuniqueness tied to the parity of r and c. For instance, the table

3	2	1
1	3	2
2	1	3

has two distinct LAD fits, with residuals and effects

0	0	-3	1
0	3	0	-1
0	0	0	0
0	-1	1	2

and

2	0	0	-1
-1	0	0	0
0	-2	1	0
0	1	0	2

respectively. We know of no general results about the nature or extent of this nonuniqueness.

4.6 Notes

1. Emerson and Hoaglin (1983) discuss the relationship between LAD fitting and median polish. They also discuss "polishing" tables using other measures of location.

2. Kemperman (1983) discusses fitting to tables by other discrete L_p criteria, $p > 1$, but focuses mainly on the case $p = 1$. He also discusses general ideas in the construction of an algorithm like median polish that would be guaranteed to converge to a LAD fit.

3. Siegel (1983) has a graph-theoretic construction for moving from a polished table to a LAD fit by forcing certain extra residuals to zero.

4. Theorem 4.2 is new, as is the procedure, modified median polish, to which it refers.

5. LAD SPLINE FITTING

5.1 Spline Functions

Polynomial spline functions are useful in numerical approximation and smoothing. Their use for smoothing of statistical data was first described by Schoenberg (1964). Several authors have pursued these ideas [e.g. Reinsch (1967, 1971); Wahba (1976); Utreras (1981b)]. The use of splines to be described in this chapter is closely related to the robust splines described by Huber (1979). Utreras (1981a) has discussed a similar problem, also using the term "robust splines", though somewhat inappropriately for statisticians, since his work is centered around the use of the discrete L_∞ norm.

> Definition 1: The function $s:R \longrightarrow R$ is a p-th degree spline function with knots at $x_1 < x_2 < \dots < x_n$ if s has p-1 continuous derivatives and a piecewise constant p-th derivative with discontinuities only at $x_1, x_2, \dots, x_n$.

This implies that between each adjacent pair of knots s is a polynomial of degree $\leq$ p, and these segments of polynomials are spliced together so that their first p−1 derivatives match at each knot. The simplest spline functions are step functions (p = 0) and continuous piecewise linear functions (p = 1). The most widely used splines apart from these are cubic splines (p = 3).

Spline functions are important because they can be flexible without using high degree polynomials. Their piecewise polynomial behavior is maintained under linear combinations, and hence the set of spline functions of a given degree and with a given set of knots is a linear space. Furthermore, the dimension of the space increases with the number of knots. The polynomials have a total of $(n + 1)(p + 1)$ coefficients, but there are np continuity constraints, and usually some boundary conditions.

There are many optimization problems whose solutions can be shown to be spline functions. Typically, these problems involve finding the smoothest function that satisfies some conditions. For example, the interpolation problem consists of finding a smooth function that interpolates a given set of data. Suppose that we agree to measure the roughness or lack of smoothness, of a function f defined on the interval [a,b] by

$$W_k(f) = \int_a^b [f^{(k)}(x)^2]\,dx \,,$$

$f^{(k)}$ denoting the kth derivative of f.

The following theorem states that the smoothest function that interpolates the data is a natural spline, where we use

Definition 2: The $(2k - 1)$ – th degree spline function s is a natural spline if

$$s^{(j)}(x_1^-) = s^{(j)}(x_n^+) = 0, \quad j \geq k.$$

This definition implies that s is a polynomial of degree $\leq k - 1$ in the interval $(-\infty, x_1)$, and another polynomial of degree $\leq k - 1$ in the interval (x_n, ∞).

Theorem 1: Given $n \geq k$ knots $x_1 < x_2 < \ldots < x_n$ in the interval (a,b) and corresponding data values $y_1, y_2, \ldots, y_n$, consider the set of functions with square-integrable k-th derivatives that satisfy the interpolation conditions

$$(1) \qquad f(x_i) = y_i \ , \ i = 1,2,\ldots,n.$$

The minimizer of $W_k(f)$ on this set is the unique natural spline of degree $(2k - 1)$ that satisfies equations (1).

Interpolating natural splines are thus the smoothest functions (as measured by W_k) interpolating the data. This interpretation is especially compelling in the case $k = 2$, where the integral is related to the energy stored in an infinitesimally bent beam. The draughtsman's tool is a cubic spline because of minimization of energy.

That there is a unique natural spline that interpolates the given set of data is made plausible by the fact that a natural spline satisfies $2k = p+1$ boundary conditions. Thus the space of natural splines with a given set of n knots has dimension n. It is easily shown that the interpolation conditions are linearly independent.

The theorem can be used to show that the solutions to a wide class of optimization problems are natural splines. Examples are

$$\text{minimize } W_k(f)$$

subject to constraints on $f(x_1), f(x_2), \ldots, f(x_n)$ and

$$\text{minimize } W_k(f) + \text{some function of } f(x_1), \ldots, f(x_n).$$

In each case we can eliminate any function not a natural spline from the search, on the grounds that there is a natural spline with the same values at $x_1, x_2, \ldots, x_n$, with a smaller value of $W_k(f)$.

The first problem of this sort to be considered led Schoenberg (1964) to discrete L_2 smoothing splines, which solve the problem of finding $f(\cdot)$ to

$$\text{minimize } W_k(f) + \alpha \sum_{i=1}^{n} \{y_i - f(x_i)\}^2.$$

There are, however, many related problems, such as [Laurent, (1972)]

$$\text{minimize } W_k(f)$$

$$\text{subject to } \max_{i=1,\ldots,n} |y_i - f(x_i)| \leq R$$

and

$$\text{minimize } W_k(f) + \alpha \sum_{i=1}^{n} |y_i - f(x_i)|.$$

The solution to the former problem is a L_∞ spline, while the solution to the latter is called a LAD spline.

A LAD spline may be regarded as a function that is fairly smooth and yet stays fairly close to the data. The parameter α determines the relative weight attached to these generally conflicting goals. As $\alpha \to 0$, the problem becomes that of minimizing $\sum |y_i - f(x_i)|$ subject to $\int [f^{(k)}(x)^2] dx = 0$. The latter constraint forces $f(\cdot)$ to be a polynomial of degree $< k$, and thus the solution is the LAD polynomial of degree k. As $\alpha \to \infty$, by contrast, the problem becomes that of minimizing $\int [f^{(k)}(x)^2] dx$ subject to $\sum |y_i - [f(x_i)| = 0$. Now the constraint forces $f(\cdot)$ to interpolate the data, and thus the solution is the interpolating natural spline of degree $(2k - 1)$. For intermediate values of α, the solution is a compromise between the two desired characteristics smoothness, and fidelity to the data.

It would also have been possible to define LAD splines in the constrained form,

$$\text{minimize } W_k(f)$$

$$\text{subject to } \sum_{i=1}^{n} |y_i - f(x_i)| \leq R.$$

However, solving the constrained problem by Lagrangian multipliers essentially consists of repeatedly solving the problem

$$\text{minimize } W_k(f) + \alpha \sum_{i=1}^{n} |y_i - f(x_i)|,$$

varying α to satisfy the constraint. Thus we regard the latter problem as the more natural, and have defined LAD splines accordingly.

We shall see in the next two sections that we also need to consider the solution to a more general optimization problem than the one that leads to LAD splines. We shall call a <u>generalized LAD spline</u> the solution to the problem

$$\text{minimize } W_k(f) + \alpha \sum_{i=1}^{n} 2c_p(y_i - f(x_i)),$$

where, for $0 < p < 1$, c_p is the <u>check</u> function,

$$c_p(x) = \begin{cases} p|x|, & x \le 0 \\ (1-p)\,|x|, & x > 0 \end{cases}$$

Note that $2c_{1/2}(x) = |x|$, and indeed, $2c_p(x) = |x| + (2p-1)x$. The piecewise linearity of the check function preserves the essential character of the optimization problem. For reasons to be discussed in the next two sections, the generalized LAD spline is also called a <u>quantile</u> spline.

5.2 Conditional and Local Quantiles

Koenker and Bassett (1978) discuss the LAD regression surface through a set of data as a generalization of the notion of the median of a single sample. They also show how other quantiles may be generalized, by minimizing a criterion similar to the sum of absolute deviations, but based on the check function. Their "regression quantiles" may also be interpreted as estimates of the conditional quantiles of Y given X, under the assumption that these quantiles are linear functions of X. The extension to nonlinear functions such as polynomials of specified degree is easy. In this section we show how a nonparametric approach may be taken.

Suppose that X and Y are random variables with a joint distribution $F(\cdot,\cdot)$, and conditional distributions described by $G(x)$, the marginal distribution function of X, and $H(y;x)$, the conditional distribution function of Y given $X = x$.

Consider the solution $m_{\alpha}(\cdot)$ to the problem

$$\begin{aligned} \text{minimize } & W_k(f) + \alpha E|Y - f(X)| \\ &= W_k(f) + \alpha \int\int |y - f(x)| \, dH(y;x) \, dG(x). \end{aligned}$$

The limit $m_{\infty}(\cdot)$, if there is one, solves the hierarchical problem

$$\text{minimize } W_k(f)$$

$$\text{subject to } \int\int |y - f(x)| \, dH(y;x) \, dG(x) = \text{minimum.}$$

Now the conditional median

$$m(x;F) = H^{-1}(1/2;x)$$

minimizes the inner integral pointwise for each x, and hence $m_\infty(\cdot) = m(\cdot,F)$, provided $m^{(k)}(\cdot,F)$ is square-integrable. Of course, $m(\cdot;F)$ is defined only on the range of X, and if this is not (a,b), we must assume that an appropriate extension exists.

This approach may be modified as follows to yield other conditional quantiles. Let

$$q_p(x;F) = H^{-1}(p;x), \; 0 < p < 1,$$

the conditional p-quantile of Y given X = x. Then $q_p(x;F)$ minimizes with respect to f the integral

$$\int c_p(y-f(x)) \, dH(y;x),$$

where c_p is the check function defined in the previous section. Hence $q_p(\cdot;F)$ is the solution to

$$\text{minimize } W_k(f)$$

$$\text{subject to } \int\int c_p(y-f(x))\, dH(y;x)\, dG(x) = \text{minimum}$$

provided it is smooth enough for the first integral to exist. If we denote the solution to

$$\text{minimize } W_k(f) + \alpha \int\int c_p(y-f(x))\, dH(y;x))\, dG(x)$$

by $q_{p,\alpha}(x;F)$, then $q_p(\cdot;F)$ is also the limit $q_{p,\infty}(\cdot;F)$.

The question now arises of how to estimate these functions given a random sample of (X,Y) pairs. One approach of course would be to assume a parametric model for the joint distribution of X and Y, and estimate the parameters and all conditional quantiles by conventional means. However, the above considerations suggest the use of $q_{p,\alpha}(\cdot;F_n)$, where F_n is the empirical distribution function of the random sample. In the case that the x-values in the sample are distinct, we can order the (x,y) pairs by x, and then $q_{p,\alpha}(\cdot;F_n)$ is the solution to

$$\text{minimize } W_k(f) + (\alpha/n) \sum_{i=1}^{n} c_p(y_i - f(x_i)),$$

which is a generalized LAD spline with knots at $x_1, x_2, \ldots, x_n$. We shall call this function a quantile spline.

Quantile splines may thus be regarded as smooth estimates of conditional quantiles in the context of data sampled randomly from a bivariate distribution. They are also useful in other contexts. Suppose for example that we observe random variables $Y_1, Y_2, \ldots, Y_n$ at times $x_1, x_1, \ldots, x_n$, and that the distribution function of the general variable Y observed at time x is $H(y;x)$. Note that this situation differs from the one we considered earlier only in that the x's are no longer assumed to be random. We can again define the quantile functions $q_p(x)$, $0 < p < 1$, but now they describe the local quantile of Y at x, rather than conditional quantiles. Two examples where quantile splines are used to estimate conditional quantiles will be given in section 4.

Laurent (1972, Chapter IX) discusses the existence and uniqueness of solutions to a general class of optimization problems. It is easily shown to include a special case closely related to the problem of constructing a quantile spline, namely the constrained version of the problem:

$$\text{minimize } W_k(f)$$

$$\text{subject to } \sum_{i=1}^{n} c_p(y_i - f(x_i)) \leq R.$$

The solutions to this problem for various values of R are the same as the quantile splines for various values of α. The existence and uniqueness of the quantile spline easily follows.

It is illuminating to compare quantile splines with the corresponding least squares splines. In the first place, it is easy to see that any quantile spline is unaffected by changing the values of any y-variables, provided no data point is moved from above the spline to below it, or vice-versa. Also, moving any data point <u>across</u> the quantile changes the quantile by only a small amount. Thus, as general smoothing tools, quantile splines are relatively insensitive to the presence of outliers in y-variables. (See section 2.3 for a more detailed discussion of the effects on LAD fits of perturbations of data.) Secondly, by calculating and displaying a few quantile splines we can reveal something of the conditional distribution of one variable given the other. For instance, the median and upper and lower quartile splines can show the center, spread, and perhaps skewness of the conditional distribution. The least squares spline can show location (in the sense of the mean rather the median), but is less easily supplemented to show other aspects of the conditional distribution.

5.3 Quantile Splines

In this section we discuss approaches to solving the minimization problem that defines a quantile spline, namely

$$(1) \qquad \text{minimize } W_k(f) + \alpha \sum_{i=1}^{n} c_p(y_i - f(x_i)).$$

We know the solution to be a natural spline, and for theoretical purposes it is convenient to express a general spline $s(\cdot)$ as a linear combination of the <u>delta-splines</u> $\{\delta_i(\cdot),\ i = 1, \ldots, n\}$, which satisfy

$$\delta_i(x_j) = \begin{cases} 1 & i = j \\ 0 & \text{otherwise} \end{cases}$$

These form a basis, and we can write

$$s(x) \equiv \sum_{i=1}^{n} s(x_i)\ \delta_i(x).$$

Thus $\{s_i = s(x_i)\ i = 1, \ldots, n\}$ are the coordinates of $s(\cdot)$ with respect to this basis.

This representation of a natural spline may be used to simplify the expression for $W_k(s)$ as follows:

$$W_k\{\sum s_i\ \delta_i\ (\cdot)\} = \int_a^b \{\sum s_i\ \delta_i^{(k)}\ (x)\}^2\ dx$$

$$= \sum_{i,j} s_i\ \Delta_{i,j}\ s_j,$$

where

$$\Delta_{i,j} = \int_a^b \delta_i^{(k)}\ (x)\ \delta_j^{(k)}(x)\ dx$$

Thus the problem (1) becomes

$$\text{minimize} \sum_{i,j} s_i\ \Delta_{i,j}\ s_j + \alpha \sum_i c_p\ (y_i - s_i).$$

The presence of the quadratic term makes this an inherently more difficult problem than the linear model estimation problems discussed in the remainder of this text. Although there are n unknowns $\{s_1,\ldots, s_n\}$, as few as k of the residuals $y_i - s_i = y_i - s(x_i)$ may be zero, or as many as n, depending on the value of α. We have not used an optimization algorithm developed specifically for this class of problems. The examples discussed in the next section are cubic splines calculated using a iteratively reweighted version of the spline fitting program described by De Boor(1978), and discussed in greater detail by Zeger and Bloomfield (1982). A somewhat different algorithm could be obtained by specializing the approach discussed by Huber (1979).

Suppose first that we wished to solve the problem

$$\text{minimize} \sum_{i,j} s_i \Delta_{i,j} s_j + a \sum_i \rho (y_i - s_i),$$

where $\rho(\cdot)$ is a smooth function. We obtain by differentiation the family of equations

$$2 \sum_j \Delta_{i,j} s_j - a \rho'(y_i - s_i) = 0, \; i = 1,\dots, n.$$

If we write $\rho'(x) = \psi(x) = xw(x)$, the equations become

$$2 \sum_j \Delta_{i,j} s_j - a (y_i - s_i) w(y_i - s_i) = 0$$

If the term $w(y_i - s_i)$ were a constant, w_i, these equations would specify the solution to the weighted least squares problem

$$\text{minimize } W_k (s) + a \sum w_i (y_i - s_i)^2,$$

which may be solved by De Boor's algorithm. Thus we may solve the more general problem by iteratively solving weighted least squares problems, and then updating the weights according to $w_i = w(y_i - s_i)$. This is an example of the approach known as iteratively reweighted least squares (a mildly redundant but conventional term).

5.4 Examples

Figure 1 shows the 0.25, 0.5 and 0.75 quantile splines, calculated with $\alpha = 1$, for a set of data of lithogenic bile concentrations in female Pima Indians [Goodall (1983, p.245)]. These curves are approximations to the lower quartile, median, and upper quartile, respectively, of the distribution of bile concentration conditionally on age. All three curves are close to linear, and the general shape of the data does not suggest that this is an artifact of the fitting. (As we noted above, as $\alpha \to 0$, the curves become linear.) However, there is a slight indication that the quantiles diverge as age increases, suggesting that a linear model in which errors have a fixed distribution may be inappropriate.

Figure 2 illustrates the difficulties that can arise in interpreting point clouds with skewed distributions. The data are 100 points with independently exponentially dsitributed coordinates. The higher density of points to the left can easily mislead the eye into believing that the typical value of y is also higher to the left. The curves, again the 0.25, 0.50 and 0.75 quantile splines, go some way towards correcting this impression.

Note that the splines in Figure 2 show more roughness than those in Figure 1, even though a smaller value of $\alpha = .2$ was used. Dimensional arguments suggest that

$$[n\ \alpha\ (\text{scale of } x)^3]/(\text{scale of } y)$$

may give a better guide to the smoothness of the splines. However, there are still many open questions about their use.

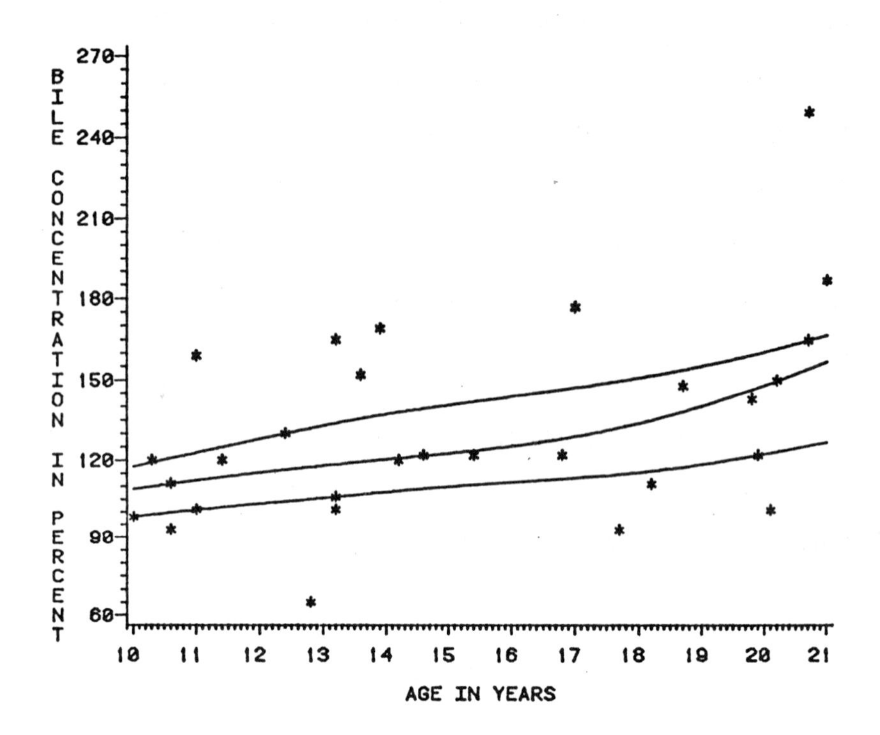

Figure 1. Bile Cholesterol Concentration in 29 Female Pima Indians, with 0.25, 0.50 and 0.75 Cubic Quantile Splines (α=1)

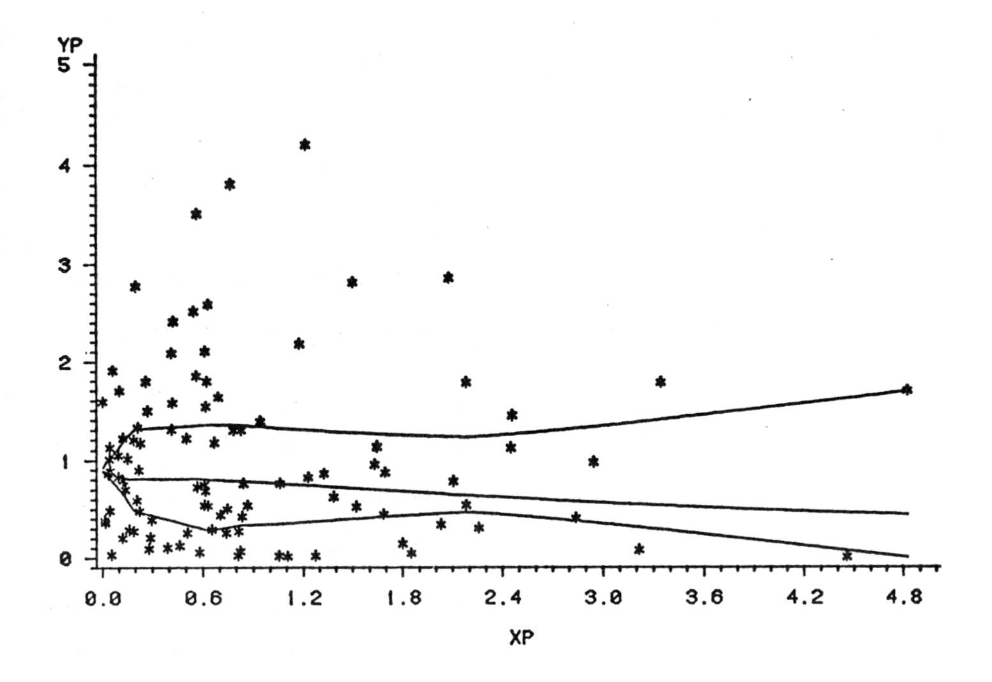

Figure 2. Random Sample of 100 Points from the Independent Bivariate Exponential Distribution, with 0.25, 0.50 and 0.75 Cubic Quantile Splines (α=0.2).

5.5 Notes

1. Theorem 1.1 is due to Schoenberg (1964).

2. Polynomial splines, especially cubic splines, have been widely discussed for solving the <u>nonparametric regression</u>

<u>problem</u> [see for example Wahba, (1974)]. Here, one observes

$$y_i = \theta(x_i) + \epsilon_i, \; i = 1,\dots,n$$

where $\theta(\cdot)$ is an unknown function to be estimated, and the ϵ_i's are independent and identically distributed errors of observation. If ϵ_i has median zero, then $\theta(x_i)$ is also the median of the distribution of y_i. Thus the nonparametric regression problem is related to the problem of finding conditional medians (in the case where the x_i's are a random sample from some distribution) or local medians (in the case where the x_i's are chosen deterministically). However, the use of quantile splines described in this chapter differs significantly from their use in nonparametric regression, in that we may explore changes in distributional <u>shape</u>, whereas in nonparametric regression the only aspect of the distribution that can change with x is the median. Philosophically this is much closer to the median traces and hinge traces constructed by Tukey (1977).

3. Cox (1983) gives some limit theorems on large sample behavior of M-type smoothing splines. However, the results do not apply exactly to quantile splines because they rest on the assumption that ρ has a bounded <u>third</u> derivative, while the <u>first</u> derivative of c_p is discontinuous. As we have remarked elsewhere, the criteria involved in LAD and related problems often do not satisfy the regularity conditions of available theorems.

4. We have not discussed the choice of α. A similar "smoothing constant" occurs in most smoothing problems. Craven and Wahba (1979) discuss the use of cross-validation to choose α for least squares smoothing splines, in the context of nonparametric regression. The similarity between smoothing splines and quantile splines suggests that cross-validation would be a very reasonable way to choose α in our context. Specifically, we would suggest finding the α that minimizes

$$\sum_{i=1}^{n} c_p(y_i - s_{p,\alpha}^{(i)}(x_i)),$$

where $s_{p,\alpha}^{(i)}(\cdot)$ is the pth quantile spline fit to the reduced set of data obtained by omitting the point (x_i, y_i). If a <u>family</u> of quantile splines were to be constructed, it would probably be wise to use a single α, found perhaps from the median spline. We note that the computational effort would be large.

5. Quantile splines are in the class of robust splines discussed by Anderssen, Bloomfield, and McNeil (1974).

6. LAD AND LINEAR PROGRAMMING

6.1 Introduction and Background

In this chapter we discuss the linear programming (LP) problem and its connection with LAD fitting. To fix the language and notation let there be given vectors $\underline{c} \in R^n$, $\underline{b} \in R^m$ and an m by n matrix A. The vector $\underline{c}$ determines a linear functional $f(\underline{x}) = \langle \underline{c}, \underline{x} \rangle$ on R^n and A and $\underline{b}$ determine m linear inequalities $A\underline{x} \le \underline{b}$. The LP problem in standard form is to

$$
\begin{aligned}
&\text{maximize } f(\underline{x}) \\
(1) \qquad &\text{subject to } A\underline{x} \le b \\
&\qquad\qquad\quad \underline{x} \ge \underline{0}
\end{aligned}
$$

The function f is called the <u>objective function</u> and $A\underline{x} \le b$ and $\underline{x} \ge 0$ are the (linear) constraints, m + n in all. The set $F = \{\underline{x} \in R^n: \underline{x} \ge 0,\ A\underline{x} \le \underline{b}\}$ is the <u>feasible region</u>, $\underline{u} \in F$ being feasible, or a feasible solution. If F is not empty, the <u>problem is feasible</u>. An <u>optimal solution</u> is $\underline{x} \in F$: $f(\underline{x}) \ge f(\underline{z})$ for all $\underline{z} \in F$. If there is an optimal solution, the problem is bounded. If all optimal solutions $\underline{x}$ satisfy $\| \underline{x} \| \leq T$ for some $T > 0$ and norm $\| \quad \|$, the solutions are bounded. By the linearity of f, the problem is bounded if F is.

The problem (1) may be expressed in many alternate forms. For example "maximizing f" is the same as "minimizing −f", and $A\underline{x} \leq \underline{b}$ is equivalent to $-A\underline{x} \geq -\underline{b}$. Another common alteration is to turn $A\underline{x} \leq \underline{b}$ into equality constraints by adding a vector $\underline{v} \in R^m$, of slack variables. Thus, defining $\underline{v} = \underline{b} - A\underline{x} \geq \underline{0}$, the constraints in (1) become

$$\begin{aligned} A\underline{x} + \underline{v} &= \underline{b} \\ \underline{x} &\geq 0 \end{aligned} \tag{2}$$

a system of m linear equations in m + n unknowns. Its feasible region is $\{(\underline{x},\underline{v}) \in R^{m+n}: A\underline{x} + \underline{v} = \underline{b}\}$, an (at least) n dimensional hyperplane in R^{m+n} intersected with the positive orthant in R^{m+n}.

The simplex method is an iterative procedure for solving (1), or deciding that it is unbounded or infeasible. In (2) write $A^* = (A|I)$, where I is the m by m identity matrix, and write $\underline{x}^* = (\underline{x},\underline{v}) \in R^{m+n}$, a vector whose first n components are $\underline{x}$ and the rest, $\underline{v}$. We then have

(3) $\underline{A}^*\underline{x}^* = \underline{b}$;

these are m linear equations in m+n unknowns. Suppose we have a non-negative solution to (3) in which n of the x_i^* are zero. In the simplest case in (2), $\underline{x} = \underline{0}$ and $\underline{v} = \underline{b}$ will do, as long as $\underline{b} \geq \underline{0}$. In general in (3), $x_{i_j}^* = 0$, $j = 1,\ldots,n$. These are the non-basic variables. The rest are basic and the basis is a set of m values, $\mathbf{B} = \{1,\ldots,m+n\}\backslash\mathbf{N}$, $\mathbf{N} = \{i_1,\ldots,i_n\}$. The m basic variables $\underline{w}$ solve the m equation system $B\underline{w} = \underline{b}$, B being the matrix comprising the columns of A^* indexed by $\mathbf{B}$, and the solution vector $\underline{x}^*$ is a basic feasible solution. Phase 1 of the simplex algorithm produces such a basic solution or else terminates with the information that the problem is infeasible.

One of the main theorems in LP theory asserts that if (1) is feasible and bounded, there is a basic, optimal solution. (Recall Theorem 1.1.1). This observation motivates Phase 2 of the simplex method.

The fundamental step moves from a basic, feasible solution $\underline{x}$ to a neighboring basic feasible solution $\underline{z}$ with $f(\underline{x}) \leq f(\underline{z})$, or stops with $\underline{x}$ as the optimal solution if no such $\underline{z}$ exists. Neighboring basic feasible solutions have m−1 common elements in their bases. Thus a non-basic variable $i \in \mathbf{N}$ is chosen to enter the basis. Then, a basic variable $j \in \mathbf{B}$ is chosen to leave.

Let $i \in \mathbf{N}$ be a candidate to enter the basis. Then there is a vector $\underline{w}_i \neq \underline{0} \in R^{n+m}$ orthogonal to the row space of A^* (dimension $\leq$ m) and to the n−1 coordinate vectors corresponding to $j \in \mathbf{N}$, $j \neq i$ (thus the jth component of $\underline{w}_i$ is zero, $j \in \mathbf{N}$, $j \neq i$). For any $t \in R$,

$$(4) \qquad \underline{x}(t) = \underline{x}' + t\, \underline{w}_i$$

also solves (3), and has zero in the jth component, $j \in \mathbf{N}$, $j \neq i$. Also $\underline{x}(0) \geq \underline{0}$ because the current solution is feasible. For each non zero component of $\underline{w}_i$ there is a value of t which makes that component of $\underline{x}(t)$ equal to zero. Let t_0 be the minimum magnitude of those values, and suppose it occurs in the jth component. If $f(\underline{x}(t_0)) \leq f(\underline{x}(0))$, variable i cannot enter the basis; its neighboring basic feasible solution $\underline{x}(t_0)$ doesn't improve the objective function. Otherwise i and j can be interchanged: $\mathbf{N} \leftarrow (\mathbf{N}\setminus\{i\}) \cup \{j\}$ and $\mathbf{B} \leftarrow (\mathbf{B}\setminus\{j\}) \cup \{i\}$.

If no $i \in \mathbf{N}$ can enter, the current solution $\underline{x}^*$ is optimal. Otherwise an element $i \in \mathbf{N}$ is chosen in a heuristic fashion to enter $\mathbf{B}$. Once done, the optimal $j \in \mathbf{B}$ to leave is easily determined as the variable corresponding to that component of $\underline{x}(t)$ in (4) which becomes zero for the smallest value of $|t|$.

The final and key step in a simplex iteration is to actually make the basis change by updating the data structure in which the problem is represented. This is accomplished by pivoting. Initially the data structure is $(A^*|\underline{b})$ in (3), an m row by m+n+1 column matrix, each row having at least one 1 in one of the distinct basis columns $j_1,\ldots,j_m$. If variable i is to replace variable j in the basis, the jth row is divided by a^*_{ij} so $a^*_{ij} \leftarrow 1$. Then, the ith column of each row $k \neq j$ is set to zero by adding an appropriate multiple of the new j^{th} row.

The LP problem (1) is important in linear optimization theory. Many combinatorial optimization problems may be cast as LP problems making it perhaps, the canonical optimization problem. Besides practical applications it encompasses an attractive theory with a variety of far-ranging aspects. Above all, it has the simplex algorithm in which all these elements combine.

The simplex method and variants have been studied extensively, perhaps more than most other algorithms. One reason probably relates to on the mystery as to why it seems to work so well. Another explanation is that it is so heavily utilized: it has been said that a non-negligible percent of ALL the computer effort expended is actually devoted to solving large linear programs!

At first glance the LAD problem only superficially resembles (1). One seeks a minimizer of $g(\underline{c}) = \sum|y_i - <\underline{c},\underline{x}_i>|$, a convex, piecewise linear function on R^k. There seems to be no feasibility aspect since the optimization is unconstrained, ranging over all R^k. Yet the argument used in proving Theorem 1.1.1 closely resembles the elementary simplex step. Both move along a line through the current (basic, feasible) point to another while the objective function improves.

In fact bounded, feasible LP problems and LAD fitting are quite intimately related. In a sense, they are equivalent, as will be shown in the next section, along with some consequences of this fact.

There are several important implications of the equivalence, some theoretical, and some algorithmic. The simplex-flavor of most LAD algorithms is a key observation that will arise frequently in Chapter 7. On the other hand, Section 3 of the current chapter contains material tending in the opposite direction, for there is potential advantage in using LAD algorithms for LP problems.

6.2 LP is LAD is LP

Suppose n points $(\underline{x}_i, y_i) \in R^{k+1}$ are given, the object being to minimize

$$(1) \qquad f(\underline{c}) = \sum_{i=1}^{n} |y_i - \langle \underline{c}, \underline{x}_i \rangle|.$$

Consider the LP problem, not in standard form,

$$(2) \qquad \text{minimize} \quad \sum_{i=1}^{n} r_i \qquad \text{subject to} \quad \begin{cases} r_i \geq y_i - \sum_{j=1}^{k} c_j x_{ij} \\ r_i \geq \sum_{j=1}^{k} c_j x_{ij} - y_i \end{cases} \quad i = 1,\dots,n.$$

It has n+k variables $r_1,\dots,r_n$, $c_1,\dots,c_k$ and 2n inequality constraints. Clearly $r_i \geq |y_i - \langle \underline{c}, \underline{x}_i \rangle|$, so the constraints in (2) are implicitly non-negativity constraints. Also, unless $y_i = \sum c_j x_{ij}$ one of the two constraints on r_i is redundant, depending on whether y_i is greater or less than $\langle \underline{c}, \underline{x}_i \rangle$.

Let $\underline{c}$ be a minimizer of f in (1). This fixes $y_i - \langle \underline{c}, \underline{x}_i \rangle$ for all i. For any choice of $r_i \geq |y_i - \langle \underline{c}, \underline{x}_i \rangle|$, $\underline{r}$ and $\underline{c}$ satisfy the constraints in (2) and $\sum r_i \geq f(\underline{c})$. Furthermore if we take $r_i = |y_i - \langle\underline{c},\underline{x}_i\rangle|$, $\sum r_i = f(\underline{c})$. This shows that the optimal solution to (2) must have $\sum r_i \leq f(\underline{c})$, the minimum of f in (1). Conversely if $r_1,\ldots,r_n$, $c_1,\ldots,c_k$ solves (2), we must have each $r_i = |y_i - \langle\underline{c},\underline{x}_i\rangle|$. If $\underline{c}$ is not a minimizer of (1) there exists $\underline{d}$ with $f(\underline{d}) < f(\underline{c})$. But then choosing $r_i = |y_i - \langle\underline{d},\underline{x}_i\rangle|$ would give a smaller value of the objective function in (2). Hence each solution of (2) provides a $\underline{c} \in R^k$ to minimize f.

The data of (1) may be described by an n by k+1 matrix, k columns for $\underline{x}$ and 1 for $\underline{y}$. In comparison if we add 2n non-negative slack variables $\underline{s}$, $\underline{t} \in R^n$ to the constraints in (2) so they become equations, we obtain

$$\text{(3)} \qquad \begin{aligned} -\sum_{j=1}^{k} c_j x_{ij} - r_i + s_i &= -y_i \\ \sum_{j=1}^{k} c_j x_{ij} - r_i + t_i &= y_i \end{aligned} \qquad i = 1,\ldots,n$$

which implies a data structure of 2n rows and 2n+k columns. To reduce the factor of 2 + 4n/k by which (2) exceeds (1) in size, the next step is to transform (2) to an equivalent LP problem.

By the duality theorem of linear programming [Chvatal (1983), for example], (2) is equivalent (solutions of one give the solutions of the other) to the LP problem

$$\text{maximize} \quad [\sum_{i=1}^{n} y_i u_i - \sum_{i=1}^{n} y_i v_i]$$

$$(4) \qquad \text{subject to} \begin{cases} u_i + v_i = 1, \; i = 1,\dots,n \\ \sum_{i=1}^{n} x_{ij} u_i - \sum_{i=1}^{n} x_{ij} v_i = 0, \; j = 1,\dots,k \\ u_i \geq 0, \quad i = 1,\dots,n \\ v_i \geq 0 \end{cases}$$

Here the dual variables u_i, v_i each correspond to the ith of the inequalities in (2). The coefficients y_i (of u_i) and $-y_i$ (of v_i) in the dual objective function in (4) are the right hand sides of the corresponding (i^{th}) inequality in (2). Finally the right hand sides, n 1's and k 0's, are the coefficients of the corresponding variables of the objective function in (2).

The size of a data structure that could hold the information in (4) is n+k by 2n. It may be viewed as the problem (2) transposed. The crucial step now consists of identifying v_i as a slack variable in $u_i + v_i = 1$. Thus, writing $v_i = 1-u_i$, (4) becomes

$$\text{maximize} \quad \sum_{i=1}^{n} y_i u_i$$

(5)

$$\text{subject to} \begin{cases} \sum_{i=1}^{n} x_{ij} u_i = \sum_{i=1}^{n} x_{ij}/2, & j = 1,\dots,k \\ 0 \le u_i \le 1, & i = 1,\dots,n \end{cases}$$

The constraint $0 \le u_i \le 1$ places u_i in the interval [0,1]. Problems like (5) are sometimes called interval programming problems (Robers and Ben-Israel) or bounded variable LP's.

There is a feature of (5) that characterizes it as an interval programming problem that arose from a LAD fit (1). In each equation, the value on the right-hand side is

$$[\sum_{i=1}^{n} x_{ij}]/2,$$

one half the sum of the left-hand side coefficients. This property turns out to be important in equating LAD fits with LP problems.

So far we have shown that to every LAD fitting problem, (1), there corresponds a bounded variable, dual LP problem of characteristic form (5). Now, a partial converse will emerge. Specifically, suppose one is given the LP problem

$$\text{maximize} \quad \sum_{i=1}^{N} d_i x_i$$

(6)

$$\text{subject to} \begin{cases} \sum_{j=1}^{N} a_{ij} x_j = b_i, & i = 1,\dots,m < N \\ x_j \geq 0, & j = 1,\dots,N \end{cases}$$

It may be a standard form problem, (1.1) to which slack variables have been added, but this is not necessary. We transform (6) to an equivalent LP of characteristic form as in (5).

The only requirement is that (6) have bounded, feasible solutions. If so, there is $T>0$ such that any optimal solution satisfies $|x_i| \leq T$. Writing $z_i = x_i/T$ in (6), the problem

$$\text{maximize} \quad \sum_{i=1}^{N} d_i z_i$$

(7)

$$\text{subject to} \begin{cases} \sum_{j=1}^{N} a_{ij} z_j = b_i/T, & i = 1,\dots,m \\ 0 \leq z_j \leq 1, & j = 1,\dots,N \end{cases}$$

is equivalent to (6) in the sense that $T\underline{z}$ gives an optimal solution of (6) if $\underline{z}$ solves (7).

The final step is to incorporate a new variable z_{N+1} to the problem in such a way as to leave unaffected any optimal solution of (7). Let K be its coefficient in the objective function and t_i its coefficient in the ith equation, in (7). Thus

$$\text{maximize} \quad Kz_{N+1} + \sum_{i=1}^{N} d_i z_i$$

$$(8) \qquad \text{subject to} \begin{cases} \sum_{j=1}^{N} a_{ij} z_j + t_i z_{N+1} = b_i/T, \ i = 1,\dots,m \\ 0 \le z_j \le 1, \quad j = 1,\dots,N+1 \end{cases}$$

If $K > 0$ is large enough z_{N+1} will be forced to 1. That would imply that each $t_i = 0$ if the N equations of (8) represent the same feasible region as that implied by (7). On the other hand if $K < 0$ is negative enough, z_{N+1} will be forced to 0 in the optimal solution. Thus, no matter what value is chosen for t_i, (8) has the same feasible region as (7). In this case if we take

$$(9) \qquad t_i = 2b_i/T - \sum_{j=1}^{N} a_{ij}$$

in (8) it will be in characteristic form: the right-hand side coefficient will equal one half the sum of left-hand size coefficients.

From this observation it is an easy matter to show that the given LP in (6) is equivalent to the LAD fit for the points $\underline{P}_i = (\underline{u}_i, v_i)$, $i = 1,\dots,N+1$ where, for $i = 1,\dots,N$, $u_{ij} = a_{ji}$, $j = 1,\dots,m$, and $v_i = d_i$; $u_{N+1,\ j} = t_j$, $j = 1,\dots,m$ and $v_{N+1} = K$. The demonstration simply consists of writing down the interval programming problem that arises from the LAD fit for the $\underline{P}_i$. It turns out to be precisely (8), from which the solutions to (6) may be obtained.

The following diagram may help clarify the relationship between LAD curve fitting and bounded, feasible LP problems.

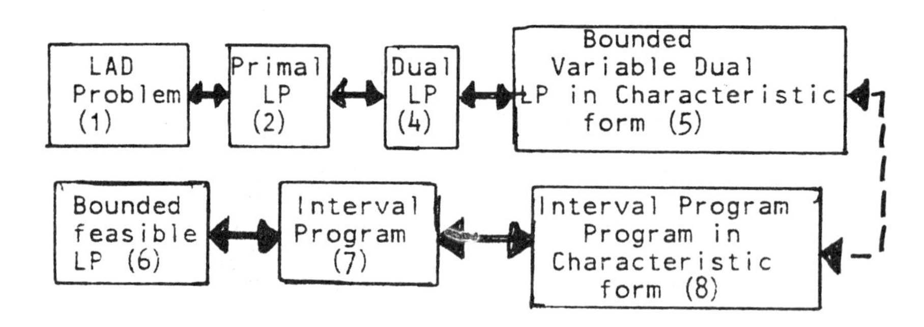

The first line describes the (reversible) derivation of the characteristic form interval program from a given LAD fit. The second describes the process of expressing a given bounded, feasible LP as a characteristic, interval program. The dotted line indicates the connection stated by the following.

Theorem 1: Any LAD curve-fit may be expressed as an equivalent bounded, feasible LP problem, and conversely.

A crucial point is that the equivalent problems are essentially of the same size. Given a LAD problem for n points in R^{k+1}, the equivalent interval programming LP has n variables and k equality constraints. Given a bounded, feasible LP with N variables and m < N equality constraints, the equivalent LAD fit is for N+1 points in R^{m+1}.

Given the LP in (6) suppose $\underline{c}$ is the equivalent LAD fit, the minimizer of

$$(10) \qquad \sum_{i=1}^{N} |d_i - \sum_{j=1}^{m} c_j a_{ji}| + |K - \sum_{j=1}^{m} c_j t_j|$$

From the preceding diagram, (6) is equivalent to the bounded-variable dual program arising from (10). Therefore the solutions $x_1,\dots,x_N$ of (6) may be easily determined using the duality theory of linear programming. Complementary slackness implies that $x_i = 0$ if $d_i \neq \sum_{j=1}^{m} c_j a_{ji}$ and $x_{N+1} = 0$ if $K \neq \sum_{j=1}^{m} c_j t_j$.

Choosing K large and negative guarantees that $x_{N+1} = 0$ so the latter relation holds. Therefore, since at least m terms in (10) are zero, N−m of the x_i in (6) are zero. Once again applying duality theory, the other x_i are directional derivatives of the function of $\underline{c}$ in (10). They are ordinarily quantities that are at hand at termination of algorithms for LAD fits, and would require no extra computations.

6.3 Some Complexity Questions

In this section we address the question of how inherently complex the LAD curve-fitting problem, (2.1) and the LP problem (2.6) are. Also of interest is the question of how hard particular algorithms for these problems will have to work, both in the worst case, and, in some sense, on the average. Since more is known about the complexity of linear programming and the simplex algorithm, Theorem 2.1 provides the main tool for statements about LAD.

Khachian (1979) created enormous surprise and interest when he showed that the LP problem is polynomial. He used an ellipsoidal shrinking method to obtain a result on linear inequalities which allowed Gacs and Lovasz(1981) to prove that the problem (2.6) of size mN may be solved in no more than f(mN) steps, f being a fourth degree polynomial.

By Theorem 2.1, the same statement must hold for LAD fits. Given n points in R^k, the LP (2.5) of size nk arises and may be solved in f(nk) steps.

The ellipsoidal shrinking idea is very important in the analysis of combinatorial optimization algorithms [Grotschel,Lovasz, and Schrijver (1980)]. However, it seems not to represent a serious choice for actually solving LP problems. Dantzig (1980a) points out most dramatically that in some average sense, the simplex method is vastly more efficient than Khachian's algorithm. For this reason we turn now to describing aspects of the complexity of the simplex method in solving LP problems, and to particular ways of performing LAD fits.

Bad problems do exist for the simplex method and its variants. Klee and Minty (1971) devised the following problem

$$\text{maximize} \quad [\sum_{i=1}^{n} 10^{n-i} x_i]$$

(1)

$$\text{subject to} \begin{cases} (2 \sum_{j=1}^{i-1} 10^{i-j} x_j) + x_i \le 100^{i-1} & i = 1,\ldots,m \\ x_i \ge 0 \end{cases}$$

Starting from an initial basic feasible solution composed of the slack variables (a conventional start), and choosing the entering variables by the steepest edge test (another common convention) the simplex algorithm would take $2^n - 1$ steps to terminate. The problem is an exponential one. This is probably not merely an artifact of the particular rule used for choosing the entering variable. Jeroslow (1973) constructed a problem of size n where, utilizing a different entering criterion – largest improvement per step – the simplex method also requires an exponential number of steps.

There is an inherent myopia in the simplex method because it only uses "local" information in choosing a non-basic variable to enter the basis. It is prohibited from "looking" beyond the next basic feasible solution in making this choice. Thus, whatever criterion is employed to optimize the current choice, there is no guarantee that ill-effects won't arise at some later iteration because of the choice just made. In fact the Klee-Minty and Jeroslow examples illustrate this principle in spades: the best step now may give rise to the worst possible over-all solution path.

Although the Jeroslow and Klee-Minty problems are exponential for a particular way of choosing the entering variable, it is not unlikely that <u>any</u> method for making basis changes will have some exponential problem to exploit its local nature. Possibly any simplex variant will have an exponential worst case.

Suppose a given LAD curve fit will be solved by applying the simplex method to the equivalent LP, (2.5). It is not clear if there are points $(\underline{x}_i, y_i) \in R^{k+1}$ such that (2.5) is of Klee–Minty, or "near" Klee–Minty form. Certainly the LP in (1) is not in the characteristic form of an LP equivalent to a LAD fit. Even more to the point, LAD fits are done by special purpose algorithms to be discussed in Chapter 7, not by using the simplex method on (2.5). Thus it is not known if there exist exponential problems for any "reasonable" LAD algorithm or in fact what the worst case behavior might be. Based upon the consequences that seem to be inherent in the blindness of one–step techniques like the simplex method, it appears likely that exponential LAD problems will exist. However in the absence of any concrete examples, the truth of the matter is still an open question.

There is some fairly persuasive evidence that worst case, or near worst case problems for variants of the simplex method are extremely rare. One kind of argument cites the behavior on thousands of real problems. On the average, the simplex method seems to require a number of iterations that is a multiple of the number of constraints.

Monte–Carlo experiments lend further support to the assertion that simplex is linear in the number of constraints. Kuhn and Quandt (1963) and Avis and Chvatal (1978) have set the simplex method on randomly generated problems. In the latter study, for a given value of m and n, the problem

$$
(2) \qquad \begin{aligned} &\text{maximize} \quad x_1 + \ldots + x_n \\ &\text{subject to} \quad \begin{cases} \sum_{j=1}^{n} a_{ij} x_j \le 1000, & i = 1,\ldots,m \\ x_j \ge 0 & j = 1,\ldots,n. \end{cases} \end{aligned}
$$

was considered. The a_{ij} are integers, randomly chosen from $\{1,\ldots,1000\}$. Whatever the choices, (2) is bounded and feasible and an initial basis of slack variables is feasible. In each of 100 repetitions, a different random realization of (2) was generated and solved. The number of simplex steps was recorded and then averaged over the 100 repetitions. For n=50 variables and m=10, 20, 30, 40, and 50 inequalities, the average number of simplex steps was 20.2, 41.5, 62.9, 78.7 and 95.2 iterations, respectively. This is evidence for a strong linear relationship of computational complexity versus m, the number of inequalities. Similar results were shown for other values of n, agreeing with the conclusion made in the Kuhn and Quandt study.

Finally, there are also some theoretical grounds for supposing that the simplex method may have an average cost that varies linearly with the number of constraints. Dantzig (1980b), assuming a certain family of distributions for the columns of A in the constraints $A\underline{x} \le \underline{b}$, showed that the expected number of simplex steps is bounded by a linear function of the number of constraints.

However both Monte-Carlo studies and the Dantzig result depend crucially upon the particular probability distributions that govern the generation of random problems. For example in another study, Dunham, Kelly and Tolle (1977) performed Monte-Carlo experiments on the simplex method. Over their population of problems the expected cost of an n variable, m constraint problem was $a + b\, n \log n \log (m/n)$ for certain constants a and b, a complexity that is <u>sub-linear</u> in m. This opens the possibility that the cost of the simplex method may depend significantly on the problem distribution. On some large classes of problems the method might be of quadratic complexity, or possibly another polynomial function of size. It might not even be of polynomial complexity.

We turn now from the simplex method to the expected complexity in solving (2.1), assuming first that the simplex method will be applied to the equivalent LP, (2.5). In order to assert linear complexity in the dimension, k, one must be able to argue that the distribution of equivalent LP's is within the family where the simplex method has been shown to be linear This seems an unrewarding approach.

However positive evidence does exist. Anderson and

Steiger (1982), in comparing various direct LAD algorithms, have discovered that average (time) cost is linear in the dimension, k. Their study will be described in the next chapter.

In summary, Theorem 2.1 furnishes the key link between the LAD and LP problems. It may be used to bound the complexity of the LAD problem by invoking the cost of ellipsoidal shrinking in solving an equivalent LP. However it allows only vague suggestions concerning the worst or average case complexity of specific algorithms that would realistically be applied to LAD curve-fitting.

6.4 Dense LP Problems as LAD Fits

The material in this section is based on Theorem 2.1. Any bounded, feasible LP problem has an equivalent LAD curve fit of essentially the same size. We present here the results of a study that indicate a potential advantage over the simplex method in applying a special purpose LAD algorithm to the equivalent LAD fit. The advantage exists when the constraint matrix A in the inequalities $A\underline{x} \leq \underline{b}$ in (2.6) is dense, and has most of its entries not equal to zero. When A is sparse, special computational techniques are invoked which enhance the efficiency of the simplex method. The purpose of this section is to point out the potential computational advantages of LAD in the dense case and perhaps to motivate the development of sparse implementations that would maintain this advantage.

In discussing this evidence, we will mention the Bloomfield-Steiger (BS) algorithm for computing LAD fits. The results of the comparisons we shall be presenting are meant to be intelligible on their own terms, and do not depend on knowing details of the BS algorithm. A proper treatment of the algorithm will appear in Chapter 7, where LAD algorithms, per se, are studied. At that point, it may be possible to explain <u>why</u> a special purpose LAD algorithm on the equivalent LAD fit may be more efficient than the simplex method on the original, dense LP problem, but for now, the goal is the mere presentation of the phenomenon.

We used the Avis-Chvatal class of test problems

$$\text{maximize} \quad [x_1 + \ldots + x_n]$$

$$(1) \qquad \text{subject to} \begin{cases} \sum_{j=0}^{n} a_{ij} x_j \leq 10{,}000, & i = 1,\ldots,m \\ x_j \geq 0 & j = 1,\ldots,n, \end{cases}$$

where n and m < n are two given integers and the a_{ij} each the average of 10 uniformly distributed random integers in $\{1,\ldots,1000\}$. Each such random problem of size mn was solved by the one-phase simplex method starting with an initial basic feasible solution composed of the slack variables, and the number of iteration steps was recorded. Then, the equivalent LAD fit was obtained for the points in the matrix,

$$(2) \qquad \begin{pmatrix} a_{11}, \ldots, a_{m1}, & 1 \\ \vdots \qquad \vdots & \vdots \\ a_{1n}, \ldots, a_{mn}, & 1 \\ t_1, \ldots, t_m, & K \end{pmatrix}$$

each row representing a point $(\underline{u}_i, v_i) \in R^{m+1}$ Each x_i is no more than 10,000 so in (3.7), T = 10,000 may be used; from (3.9), the value $t_i = 2 - \sum_{j=1}^{n} a_{ij}$ was taken.

The BS algorithm performed the LAD fit on the points in (2) from which the solution to (1) was simply be "read off" as described in Section 2. The iteration count was recorded. This quantity was deemed to be perfectly comparable to the simplex iteration count because in both algorithms the costly manipulation performed in each iteration is the pivot operation on an m by n matrix (as will be seen in Chapter 7, the BS and other algorithms pivot on the columns of the data matrix).

Ten random problems (1) of size mn were generated,

then solved by both methods. The iteration counts were then averaged over the 10 repetitions. The results are summarized in Table 1.

TABLE 1

Comparison of LAD and the one-phase simplex method in solving the LP problem (1), iteration counts averaged over 10 independent random instances.

		n = 5	10	20	50	
	5	4.6	6.9	8.4	9.9	SIMP
		3.1	4.0	5.1	6.7	LAD
m	10		10.8	16.3	22.6	SIMP
			7.1	10.0	12.7	LAD
	20			24.6	42.6	SIMP
				14.8	25.8	LAD

The simplex iteration counts accord closely with those reported by Avis and Chvatal. Timings have not been included due to the possibility that differences between computers, systems, and actual coding, could alter any results.

Nevertheless the timings do not contradict the Table

results. For example, using a simple FORTRAN code for the table form of the simplex method, the timings corresponding to the results presented in Table 1 give an advantage to LAD of 8% for m=5, n=10 and 30% for m=10, n=50.

In the foregoing experiments an initial feasible solution was presented to the simplex algorithm. In general, the two-phase simplex method would be required and this would force simplex to work harder on each problem because a feasible solution would have to be obtained first. LP problems for which an initial feasible solution is not immediately available would not cause any extra effort for the LAD equivalent.

Thus in a second set of experiments, problems (1) were solved by both the two-phase simplex method and as LAD equivalents. The former began with a phase-one initial basis composed of artificial variables. The results appear in Table 2.

TABLE 2

Comparison of LAD and the two-phase simplex method in solving the LP problem (1), iteration counts averaged over 10 independent random instances.

m		n = 5	10	15	20	25	30	40	50
5	SIMP	7.7	11.1	11.8	13.7	13.5	13.5	15.4	14.1
	LAD	3.2	3.9	4.8	5.3	6.4	6.4	7.2	6.8
10	SIMP		21.1	26.1	29.6	29.5	31.3	35.7	39.5
	LAD		6.4	8.0	10.2	11.0	11.5	13.7	13.3
15	SIMP			36.1	47.9	46.8	45.8	58.9	60.0
	LAD			10.1	12.6	15.0	15.6	19.0	20.4
20	SIMP				56.8	68.5	69.8	76.3	81.5
	LAD				13.9	16.1	18.5	21.3	24.8

The LAD results for the new set of problems agree closely with those shown in Table 1 and give a rough idea of the variability of performance statistics such as average iteration count. If 100 instances had been averaged, the variability between tables would have been reduced.

Clearly less work is done by BS on the LAD equivalent than by the two phase simplex method applied directly to (1). On the larger problems, less than 1/3 as many iterations were required.

Again, while CPU timings may be misleading, timing data does not contradict Table 2. The LAD runs were from 1 1/4 to 3 times faster than the two phase simplex runs.

In the next chapter we discuss LAD algorithms in detail. There, some aspects of BS that may explain its relative efficiency will emerge. For the present we simply present the computational results. They support a continued interest in solving LP problems via LAD curve fitting. In general, it would be interesting to understand – in terms of simplex steps on a given LP problem – what a single iteration of, say, BS does on the equivalent LAD fit. This could possibly help explain the apparent efficacy of the latter algorithm. Secondly, there is the possibility that the ideas in BS, if implemented in a way in which sparsity might be exploited, could be useful in large LP codes.

6.5 Notes

1. See Chvatal (1983) for information on the theory, applications, and algorithmic aspects of LP. It is a comprehensive and elegant source.

2. The problem (1.1) may be traced at least as far back as Fourier (1824), who studied properties of linear inequality systems [see Kohler (1973)].

3. Dantzig (1947) is generally credited with the

development of the simplex algorithm although Kantorovitch (1939) proposed a quite similar method for a restricted class of problems.

4. The equivalence of (2.1) and (2.2) was pointed out in Charnes, Cooper, and Ferguson (1955) but Harris (1950) was apparently the first to state that LAD fits could be accomplished by setting up, then solving, an appropriate LP.

5. The derivation of (2.5) from (2.2) is due to Wagner (1959). This bounded variable dual form of the LP equivalent to the LAD fitting problem was used for at least ten years as the basis of computer programs. It was called interval programming by Robers and Ben-Israel (1969).

6. The converse part of Theorem 2.1 is apparently due to Witzgall (preprint). It is mentioned in Bartels, Conn, and Sinclair (1978).

7. The computational cost of converting the solution of (2.10) to that of (2.6) is negligible. The non-zero x_i's are directional derivatives of the objective function. These quantities are usually available at termination of most LAD algorithms.

8. Ellipsoidal shrinking is actually due to Schor (1977) and others.

9. Jeroslow (1973) actually shows that there is an exponential example, even if the simplex method is allowed to look ahead a fixed number, k, of steps.

10. Bartels (1980) has a non-simplex algorithm that seems superior to variants of the simplex method in average iteration counts, but the savings seem smaller than those reported in Table 4.2.

7. ALGORITHMS FOR LAD

7.1 Introduction and Background

This chapter deals with algorithmic considerations in performing LAD fits. From a historical standpoint, very little attention seems to have been directed towards LAD fitting until the 1950's. Following Edgeworth's work in the 1880's there was relative silence for many years, probably because of computational difficulties.

E.C. Rhodes (1930) and Singleton (1940) could be grouped as the first "modern" workers in the area. Rhodes showed how to fit a LAD quadratic to a set of points in the plane. His procedure iterated a sequence of recursive calls to a program which obtained linear LAD fits. This, in turn, iterated a sequence of recursive calls to a procedure for finding LAD fits through the origin, as in (1.2.1). Singleton correctly suggested that the Rhodes algorithm would easily generalize beyond quadratics but produced a different kind of recursive – and iterative – method of his own for the general case.

Interestingly, neither method was implemented once computers became available. Perhaps an explanation rests on the complexity of these procedures, but it may be that the advent of linear programming codes obviated the need for either of these two algorithms.

T. Harris (1950) suggested that

$$(1) \qquad f(\underline{c}) = \Sigma \; |y_i - \langle \underline{c}, \underline{x}_i \rangle|$$

could be minimized using linear programming theory but Charnes, Cooper, and Ferguson (1955) actually showed that the primal LP problem (6.2.2), or its dual, (6.2.4), could be solved via the simplex method and give the minimizer of (1). This was probably the birth of automated LAD fitting, the bounded variable formulation (6.2.5) of Wagner (1959) probably being the most efficient method for several years.

In the late 1960's and throughout the 1970's special purpose (LP-based) LAD algorithms were introduced with a frequency that sometimes exceeded their novelty. Some are reminiscent of rediscovering the wheel in that the authors (the present ones included) were not fully cognizant of earlier work. In other cases, new contributions were not recognized as being equivalent to existent methods. Finally in one interesting case, genuninely distinct mathematical approaches give rise to algorithms which, after a few iterations, operate within a situation where their differences have disappeared. If started appropriately they always generate identical computations!

In discussing techniques for obtaining LAD fits we will need to treat the special purpose algorithms and highlight similarities and differences. Our technique for doing this will be to focus in detail on three of the best algorithms now available, those of Barrodale-Roberts (BR), Bartels-Conn-Sinclair (BCS), and Bloomfield-Steiger (BS). They have a key feature in common: All three exploit the Gauss LP characterization in Theorem 1.1.1, that there is an optimal fit in (1) with at least k zero residuals, (assuming X is of full rank). These are extreme fits. The algorithms iterate to the optimum via a sequence of extreme fits along which the value of f in (1) is non-increasing. Each has a startup phase which terminates when the first extreme fit is found. Thereafter, successive extreme fits $\underline{c}_i$ and $\underline{c}_{i+1}$ differ only with respect to one zero residual. More precisely, if $r_j(\underline{c}_i) = 0$, $j \in \mathbf{B}_1 = \{j_1,...,j_k\}$, one element of $\mathbf{B}_1$ is replaced by one element in $\mathbf{N}_1$, the complement of $\mathbf{B}_1$ to give $\mathbf{B}_2$, and $r_j(\underline{c}_{i+1}) = 0$, $j \in \mathbf{B}_2$. Hence $|\mathbf{B}_1 \cap \mathbf{B}_2| = k-1$. The only possible differences between two algorithms that have these properties could be (a) how they startup, (b) how they choose the element $i \in \mathbf{B}$ to be replaced, and (c) how they choose the element in $\mathbf{N}$ to replace i, and we shall use this observation in the next section.

It is no accident that this discussion seems close to the flavor of the simplex method for LP problems. Indeed, most LAD algorithms, and in particular the three we shall focus on in the next section, are actually variants of the simplex method on a certain LP. Startup phases are analogous to Phase I of the simplex method and successive fits correspond to neighboring basic feasible solutions. Finally, the updating of $\mathbf{B}_1$ to $\mathbf{B}_2$ where $|\mathbf{B}_1 \cap \mathbf{B}_2| = k-1$, is analogous to change of basis in the simplex method. To make these correspondences explicit one need only identify the extreme fits in (1) with basic feasible solutions to (6.2.2). For expository purposes we prefer the former framework because it seems so much more natural to deal directly with the parameter space $\underline{c} \in R^k$ and its geometry, and to refer directly to the points $(\underline{x}_i, y_i)$ being fit. Accordingly, we shall adhere to it throughout. In doing so we rely on Section 1.3 especially Theorems 1.3.5 - 1.3.9.

Section 2 describes the BR, BCS and BS algorithms, with particular emphasis on the similarities and differences in (a) startup and (b) how the next extreme fit is obtained from the current one. The Notes section will fill in many of the gaps we leave by citing other contributions and placing them both conceptually in relation to the three main algorithms, and in the time sequence in which the contributions originally occurred. Section 7.3 contains the results of a study that compares the three algorithms (and one modification) of the previous section. Differences in performance are related to features of these algorithms. Computational complexities are studied.

In fitting $(\underline{x}_i, y_i) \in R^{k+1}$, $i = 1,\ldots,n$, the amount of work required would be expected to increase with k, the number of parameters in the fit, $n > k$ being fixed. In Section 4 we describe an idea due to Seneta and Steiger (1983) which maps (1) into a complementary problem. This approach gives rise to a class of algorithms for slightly over-determined equation systems that increase in efficiency as k increases towards n. In conjunction with the material of Chapter 6, these techniques represent a non-dual method for reducing the size and computational requirements of an LP problem.

Finally Section 7.5 presents some other methods for computing or approximating LAD fits. Amongst the latter is the interesting iteratively reweighted least squares (IRLS). Another is Clark's (1983) exact, finite algorithm for Huber M-estimators which gives LAD and LSQ as special cases. The efficacy of these methods is compared to that of the Section 2 procedures.

7.2 Three Special Purpose LAD Algorithms

We describe and compare the main aspects of the three best LAD algorithms currently available. These are due to Barrodale-Roberts (1973, 1974), Bartels-Conn-Sinclair (1978), and Bloomfield-Steiger (1980a), and are abbreviated by BR, BCS, BS, respectively.

7.2.1 The Barrodale-Roberts Algorithm

The work of Barrodale and Roberts is of primary importance, for it represents the incisive steps that enable an LP formulation of minimizing

$$(1) \qquad f(\underline{c}) = \sum_{i=1}^{n} |y_i - \langle \underline{c}, \underline{x}_i \rangle|, \ \underline{c} \in R^k$$

to be exploited efficiently. The mathematical basis of BR is to cast (1) as a primal LP problem and to cleverly modify both the simplex method and the required data structure so that the former is efficient and the latter is parsimonious. For the moment we will bypass their LP formulation of (1) and instead, describe the BR algorithm in a direct manner, relying on the terminology and results of Sections 1.2 and 1.3.

The <u>startup phase</u> has at most k steps. It begins with

$\underline{c}_0 = \underline{0} \in R^k$ and either stops with the optimum $\underline{c}_j$, $j < k$, or else with $\underline{c}_k$, the first extreme approximation encountered. In this case $r_i(\underline{c}_k) = 0$ for at least k values of i stored in a set B, and $\{\underline{x}_i, i \in B\}$ spans R^k.

Each new step will "activate" a new coordinate direction in R^k that had been restricted to be zero up to the current step, so by step k, all components of $\underline{c}_k$ will be active and may be non-zero. Furthermore, at step j, the fit $\underline{c}_j$ will be determined by a basic set B of j pointer values at which $r_i(\underline{c}_j) = 0$, $i \in B$. The basic set for $\underline{c}_{j+1}$ is $B^* = B \cup \{m\}$, $m \notin B$, consisting of one additional pointer value, and $f(\underline{c}_j) \geq f(\underline{c}_{j+1})$. Finally BR uses a steepest "edge" criterion for heuristically deciding what direction to move along from $\underline{c}_j$ to the next approximation.

In describing the specific details we maintain a vector $\underline{\tau} = (\tau_1,\ldots,\tau_k)$ of pointers to distinct elements of $\{0,1,\ldots,k\}$. Nonzero components of $\underline{\tau}$ describe which unit coordinate vectors $\underline{\epsilon}_i \in R^k$ are <u>active</u>. They define the subspace of R^k in which approximations made thus far have been restricted. Initially $\underline{\tau} = \underline{0}$.

Another vector $\underline{\sigma} = (\sigma_1,\ldots,\sigma_k)$ points to distinct elements of $\{0,1,\ldots,n\}$. Nonzero elements of $\underline{\sigma}$ describe basic vectors $\underline{x}$ which determine the current approximation $\underline{c}_j$, so $y_m = \langle \underline{c}_j, \underline{x}_m \rangle$

if $m \in B = \{i: \sigma_i \neq 0\}$, the <u>basic set</u>. Initially $\underline{\sigma} = \underline{0}$, $B = \phi$, and the <u>basis</u> consists of the k unit coordinate vectors $\underline{\epsilon}_i \in R^k$. At each startup step one coordinate vector will be replaced by a non-basic $\underline{x}_i$. Finally at each step there is a set of k linearly independent <u>"edge" directions</u> $\underline{\delta}_1,...,\underline{\delta}_k$, initially the unit coordinate vectors in R^k. At the start, $\underline{c}_0 = \underline{0}$.

For the first iteration, the directional derivatives of f at $\underline{c}_0$ in directions $\pm\,\underline{\delta}_i$ are computed using (1.3.18), to obtain

(2) $$f'(\underline{0}, \pm\,\underline{\delta}_i) = \sum_{j \in Z} |x_{ji}| \mp \sum_{j \notin Z} x_{ji} \text{ sign } (r_j(\underline{0})),$$

Z here denoting $\{j: y_j = 0\}$. By the convexity of f, either $f'(\underline{0},\underline{\delta}_i)$ and $f'(\underline{0},-\underline{\delta}_i)$, are both non-negative, or of opposite sign.

Suppose (2) is most negative along the line $t\underline{\delta}_p$, in this case the steepest "edge" emanating from $\underline{c}_o$. BR moves to its next approximation by minimizing f along $t\underline{\delta}_p$. The function

$$f(t\underline{\delta}_p) = \sum |y_i - t\langle\delta_p,\underline{x}_i\rangle|$$

is mimimized when $t = t_q = y_q/x_{qp}$, the weighted median of $\{y_i/x_{ip}\}$ with weights $|x_{ip}|$. This gives $\underline{c}_1 = t_q\underline{\delta}_p$ as the next approximation and

$$r_q(\underline{c}_1) = y_q - \langle \underline{c}_1, \underline{x}_q \rangle = 0$$

so $|\mathbf{Z}_{\underline{c}_1}| \geq 1$. Recall that $\mathbf{Z}_{\underline{c}} = \{i : r_i(\underline{c}) = 0\}$. At this point we set $\tau_1 = p$ (the pth coordinate direction is active) and $\sigma_1 = q$, so $\mathbf{B} = \{q\}$ and $\{\underline{x}_q\}$ replaces $\underline{\epsilon}_p$ in the basis.

In preparation for step 2, the edge directions $\underline{\delta}_i$ will be modified so that $\langle \underline{\delta}_p, \underline{x}_q \rangle = 1$ and $\langle \underline{\delta}_i, \underline{x}_q \rangle = 0$, $i \neq p$. This assures that all parameter values $\underline{c}_1 + t\underline{\delta}_i$ define hyperplanes which contain the current basic point $(\underline{x}_i, y_i)$; i.e.,

$$r_q(\underline{c}_1 + t\underline{\delta}_i) = y_q - \langle \underline{c}_1, \underline{x}_q \rangle - t\langle \underline{\delta}_i, \underline{x}_q \rangle = 0.$$

Thus updating by

$$(3) \qquad \begin{aligned} &\underline{\delta}_i \leftarrow \underline{\delta}_i - (x_{qi}/x_{qp})\, \underline{\delta}_p \qquad i \neq p \\ &\underline{\delta}_p \leftarrow \underline{\delta}_p / x_{qp} \end{aligned}$$

assures that the $\underline{\delta}_i$ are still linearly independent and that

$$(4) \qquad \begin{aligned} &\langle \underline{\delta}_p, \underline{x}_q \rangle = 1 \\ &\langle \underline{\delta}_i, \underline{x}_q \rangle = x_{qi} - (x_{qi}/x_{qp})\, x_{qp} = 0, \qquad i \neq p \end{aligned}$$

Step 2 may now begin.

By induction, at the start of step j+1 we have the approximation $\underline{c}_j$ determined by j distinct basis vectors $\underline{x}_{\sigma_i}$, $i = 1,\ldots,j$ at which $r_{\sigma_i}(\underline{c}_j) = y_{\sigma_i} - \langle \underline{c}_j, \underline{x}_{\sigma_i} \rangle = 0$. Also $\tau_1,\ldots,\tau_j$ point to the active coordinate directions which span that part of the parameter space visited thus far; i.e., $\underline{c}_j$ has i^{th} coordinate equal to zero if $i \neq \tau_1,\ldots,\tau_j$. Finally the independent edge direction vectors $\underline{\delta}_1,\ldots,\underline{\delta}_k$ satisfy

$$(5) \qquad \langle \underline{\delta}_i, \underline{x}_{\sigma_m} \rangle = \begin{cases} 1 \text{ if } i = \tau_m \\ 0 \text{ otherwise} \end{cases}$$

for $i = 1,\ldots,k$ and $m = 1,\ldots,j$.

To start the step, the directional derivatives of f at $\underline{c}_j$ in the directions $\pm\,\underline{\delta}_i$ are computed for each i such that $\tau_i = 0$ using

$$(6) \qquad f'(\underline{c}_j, \pm\,\underline{\delta}_i) = \sum_{m \in Z} |\langle \underline{\delta}_i, \underline{x}_m \rangle| \mp \sum_{m \notin Z} \langle \underline{\delta}_i, \underline{x}_m \rangle \operatorname{sign}(r_m(\underline{c}_j))$$

For any i, at most one can be negative, by convexity.

Suppose (6) is positive. Assuming $|Z| = j$, $(c_{\tau_1},\ldots,c_{\tau_j})$ is a non-degenerate extreme point of the sub problem of (1) restricted to the active directions determined by $\underline{\tau}$, and by

Corollary 1.3.2, it is the unique optimal point. A modification of Corollary 1.3.1 shows that $\underline{c}_j$ is optimal for (1).

Ignoring other cases where $|Z| > j$, we let $\underline{\delta}_p$ be the direction where (6) is most negative. Then $\underline{c}_{j+1} = \underline{c}_j + t\underline{\delta}_p$ where $t = t_q = (y_q - \langle \underline{c}_j, \underline{x}_q \rangle)/\langle \underline{\delta}_p, \underline{x}_q \rangle$ is the weighted median that minimizes $\sum |y_i - \langle c_j, \underline{x}_i \rangle - t\langle \underline{\delta}_p, \underline{x}_i \rangle|$, $q \notin B$. Preparing for the next step we update as follows:

$$\tau_{j+1} \leftarrow p,\ \sigma_{j+1} \leftarrow q,\ B \leftarrow B \cup \{q\}$$

$$\underline{x}_q \text{ replaces } \underline{\epsilon}_p \text{ in the basis}$$

$$(7) \qquad \underline{\delta}_i \leftarrow \underline{\delta}_i - (\langle \underline{\delta}_i, \underline{x}_q \rangle / \langle \underline{\delta}_p, \underline{x}_q \rangle)\underline{\delta}_p, \quad i \neq p$$

$$\underline{\delta}_p \leftarrow \underline{\delta}_p / \langle \delta_p, \underline{x}_q \rangle.$$

It is easy to verify that (5) now holds for $i = 1,\ldots,k$, $m = 1,\ldots, j+1$. If $m = 1,\ldots,j$, the third line of (7) still has $\langle \underline{\delta}_i, \underline{x}_q \rangle = 0$, $i \neq p$ while the last forces $\langle \underline{\delta}_p, \underline{x}_q \rangle = 1$, as required.

The startup is finished when $j+1 = k$. At this point all coordinate directions are active and $\underline{c}_k$ is an extreme fit. By (5), the direction matrix

$$D = (\underline{\delta}_{\tau_1},\ldots,\underline{\delta}_{\tau_k})$$

is the inverse of $B = X(\underline{\sigma})$, the matrix whose rows are $\underline{x}_{\sigma_1},\ldots,\underline{x}_{\sigma_k}$, so

$$(8) \qquad BD = I.$$

In step $j+1$, $j \geq k$, the startup guarantees that the fit $\underline{c}_j$ is extreme and (8) is satisfied by the matrix B of basic vectors and corresponding directions. Directional derivatives are now computed as in (6), but for all directions $\underline{\delta}_i$, $i = 1,\ldots,k$. If $\underline{c}_j$ is non-degenerate and all derivatives are non-negative, we are finished, by Corollary 1.3.2. Otherwise suppose $\underline{\delta}_{\tau_m}$ defines the steepest downhill edge. Thus, the next fit will be $\underline{c}_{j+1} = \underline{c}_j + t\underline{\delta}_{\tau_m}$, where the weighted median

$$t_q = (y_q - <\underline{c}_j,\underline{x}_q>)/<\underline{\delta}_{\tau_m},\underline{x}_q>, \ q \notin \mathbf{B}$$

is obtained by minimizing

$$\Sigma \ |y_i - <\underline{c}_j,\underline{x}_i> - t <\underline{\delta}_{\tau_m},\underline{x}_i>|$$

Moving along $\underline{\delta}_{\tau_m}$ corresponds to removing σ_m from **B**. Stopping at t_q corresponds to making $\underline{x}_q$ basic. Thus $\sigma_m \in \mathbf{B}$ is replaced by q: $\sigma_m \leftarrow q$. To update $D = (\underline{\delta}_1,\ldots,\delta_k)$ we perform (7) with $p = \tau_m$, and this preserves (8) for the new B whose new m^{th} row is $\underline{x}_q$.

A convenient data structure with which to express the algorithm is

$$(9) \qquad A = \begin{pmatrix} X & \underline{y} \\ I & \underline{0} \end{pmatrix}$$

$(X\,|\,\underline{y})$ is the n by k+1 data matrix from (1), I a k by k identity matrix, and $\underline{0} \in R^k$ a vector of zeros.

The directional derivatives in (2) are easily obtained from the first n rows of A. For column i set $g_i = \sum |a_{ji}|$ and $h_i = \sum a_{ji}$ sign $(a_{j,k+1})$, the first sum over j, $1 \le j \le n$, where $a_{j,k+1} = 0$, the second over the rest. The minimum of $f_i = \min(g_i - h_i, g_i + h_i)$, $1 \le i \le k$ locates the steepest edge, say $\underline{\delta}_p$. To update as in (3), the following pivot is performed on A. The pth column is divided by a_{qp} and then, for $i \ne p$, the ith column is modified by subtraction of a_{qi} times the new p^{th} column, $1 \le i \le k+1$, in all n+k rows. The updated data structure is now

$$(10) \qquad A_1 = \begin{pmatrix} XD_1 & \underline{r} \\ D_1 & -\underline{c}_1 \end{pmatrix}$$

where D_1 is the k by k matrix of current direction columns, XD_1 the n by k matrix of inner products $(<\underline{x}_i,\underline{\delta}_j>)$, $\underline{c}_1$ the first approximation, and $\underline{r}$ the residuals from this first fit. So σ_1 points to the basic row of A and $A_1(\sigma_1,\tau_1) = 1$, by (4).

In general, after startup, as we begin step $j + 1$, $j \geq k$ we will have

$$(11) \qquad A_j = \begin{pmatrix} XD_j & \underline{r} \\ D_j & -\underline{c}_j \end{pmatrix}$$

The k columns of D_j contain the current directions $\underline{\delta}_1,\dots,\underline{\delta}_k$, $\underline{c}_j$ is the current fit, and $r_i(\underline{c}_j) = y_i - <\underline{c}_j,\underline{x}_i>$. $XD_j = (<\underline{x}_i,\underline{\delta}_j>)$ is the matrix of inner products which, by virtue of (5), contains unit vectors in the basis rows $\sigma_1,\dots,\sigma_k$. Directional derivatives are computed from (6) by simply using the ith column entries of A_j for $<\delta_i,\underline{x}_m>$ and the $k+1^{st}$ column entries for $r_m(\underline{c}_j)$.

The BR algorithm could be summarized as follows.

Given $(\underline{x}_i,y_i) \in R^{k+1}$, $i = 1,\dots,n$.

Initialize:

[1] $A \leftarrow \begin{pmatrix} X & \underline{y} \\ I & \underline{0} \end{pmatrix}$

[2] $j \leftarrow 0;\ \underline{\sigma} \leftarrow \underline{\tau} \leftarrow \underline{0} \in R^k$

Next Descent Direction:

[3] $\mathbf{Z} \leftarrow \{i,\ 1 \le i \le n:\ a_{i,k+1} = 0\}$, $\mathbf{N}$ = the rest

[4] For m = 1 thru k DO

$g_m \leftarrow \sum_Z |a_{im}|$

$h_m \leftarrow \sum_N a_{im}\ \mathrm{sign}\ (a_{i,k+1})$

$f_m \leftarrow \min\ (g_m - h_m,\ g_m + k_m)$

END

[5] $f_p \leftarrow \min\ (f_i;\ \tau_i = 0)$ if $j < k$ (startup), or

$f_p \leftarrow \min f_i$ if $j \ge k$

[6] go to 12 if $f_p \ge 0$

New Basic Vector:

[7] $t \leftarrow a_{q,k+1}/a_{qp}$, the weighted median of

$\{a_{i,k+1}/a_{ip},\ i = 1,\dots,n\}$ with weights $|a_{ip}|$.

Update:

[8] Pivot (q,p): col $p \leftarrow$ col p/a_{qp}
For $i \neq p$, $i \leq k+1$, col $i \leftarrow$ col $i - a_{pi}$ col p.

[9] $j \leftarrow j+1$; if $j > k$ go to 11

[10] $\sigma_j \leftarrow q$; $\tau_j \leftarrow p$; go to 3

[11] $\sigma_p \leftarrow q$;
Find m: $\sigma_m = p$. Set $\sigma_m \leftarrow 0$
go to 3.

Output:

[12] $c_i \leftarrow -a_{n+i,k+1}$, $i = 1,\ldots,k$.
$\text{fmin} \leftarrow \sum_{i=1}^{n} |a_{i,k+1}|$

We now show how Barrodale and Roberts developed the algorithm from an LP formulation. They cast (1) as the primal LP problem

$$\text{minimize} \quad \sum_{i=1}^{n} (u_i + v_i) \quad .$$

$$(12) \qquad \text{subject to} \begin{cases} y_i = \langle \underline{b}, \underline{x}_i \rangle + u_i - \langle d, \underline{x}_i \rangle - v_i \\ b_j,\ d_j \geq 0,\ j = 1,\ldots,k. \\ u_i,\ v_i \geq 0,\ i = 1,\ldots,n. \end{cases}$$

Clearly $\underline{b}-\underline{d}$ represents $\underline{c}$ and $\underline{u}-\underline{v}$, the residuals, $\underline{y}-X\underline{c}$.

Two observations allow this formulation to become efficient. The first is that it is unnecessary to specifically account for both $\underline{b}$ and $\underline{d}$, since $b_j d_j = 0$ can always be arranged. Similarly $u_i v_i$ may be taken as zero.

Because of the special structure, the information for the large LP problem in (12) may be collapsed into a data set like (9). For example negative y_i indicate that those v_i are basic and initially have values $v_i = |y_i|$; non-negative y_i indicate the basic u_i's, initially equally to y_i; initially $\underline{b},\underline{d}$ are non-basic and equal $\underline{0} \in R^k$.

Barrodale and Roberts make a second observation that permits significant streamlining of the simplex method applied to (12) by reducing the number of pivot steps taken. The initial basic feasible solution is built on a basis composed of $\underline{u}$'s and $\underline{v}$'s; $u_i = y_i$ if $y_i \geq 0$, $v_i = y_i$ if $y_i < 0$. Here $\underline{c}$ and $\underline{d}$ are non-basic so $c_i = d_i = 0$. At each step a basic u_i or v_i

leaves. In Phase I it is replaced by one of the c_i or d_i while in Phase II, the non-basic u_i, v_i are the candidates to enter.

Suppose a non-basic variable, α_r, has been chosen to enter the basis. Its value will increase from zero, while maintaining the n equalities in (12). In the conventional simplex implementation, the entering variable would be increased up to that point where the first basic variable, say β_s, becomes zero. Any further increase of α_r would violate the feasibility of β_s. At this point a pivot operation would replace β_s by α_r and a new non-basic variable would be chosen, etc.

The key insight of BR is that α_r may possibly be increased further, and $\sum(u_i + v_i)$ decreased further, without pivoting out β_s. Specifically, suppose u_{s_1} (or v_{s_1}) is the basic variable to first become zero as α_r is increased to say, t_1. Since $u_{s_1} v_{s_1} = 0$ one can increase α_r beyond t_1 and still maintain the feasibility of both u_{s_1} and v_{s_1} if the roles of u_{s_1} and v_{s_1} are interchanged. Thus as α_r is increased beyond t_1, instead of having u_{s_1} (or v_{s_1}) become negative, u_{s_1} (or v_{s_1}) would be held at zero and v_{s_1} (or u_{s_1}) made to increase.

This may be continued. If $\sum(c_i + d_i)$ still decreases, α_r may be increased to say, $t_2 > t_1$ at which the next basic variable u_{s_2} (or v_{s_2}) first becomes zero. The roles of these variables are interchanged, u_{s_2} (or v_{s_2}) is held at zero and v_{s_2} (or

u_{s_2}) increases as a_r increased. As long as $\Sigma(c_i + d_i)$ decreases as a_r increases, the switching of u_{s_i} with v_{s_i} continues, and no costly pivots are required. Finally when a_r has increased to a certain point, say t_m, at which a basic variable u_{s_m} (or v_{s_m}) first becomes zero, any further increase in a_r will make $\Sigma(c_i + d_i)$ increase. At this point a_r replaces u_{s_m} (or v_{w_m}) in the basis via the pivot operation.

Barrodale-Roberts' introduction of these multiple pivot sequences accomplishes several standard simplex steps by performing only one pivot operation. If this were described in the standard terminology, it would correspond to a weighted median calculation for the line search of finding the minimum of f along a line $\underline{c} + t\underline{\delta}$.

Suppose the non-basic variable chosen to enter is u_m (or v_m) [This means we are in Phase II, or beyond the startup. A similar discussion pertains for Phase I where some b_i (or d_i) would be entering]. This implies, in the original terminology that at the current fit, $\underline{c}_j$, $y_m = \langle \underline{c}_j, \underline{x}_m \rangle$ and that m is going to leave the basic set. This corresponds to moving down one of the edge directions, for if $m = \sigma_p$, $1 \leq p \leq k$, the entire line $\underline{c}_j + t\delta$, where $\underline{\delta} = \underline{\delta}_{\tau_p}$, contains all basic points except $(\underline{x}_m, y_m)$. As t varies

$$f(\underline{c}_j + t\underline{\delta}) = \sum |y_i - \langle \underline{c}_j, \underline{x}_i \rangle - t\langle \underline{\delta}, \underline{x}_i \rangle|$$

$$= \sum |w_i - tz_i|$$

is minimized at the weighted median of the $\{w_i/z_i\}$. Its graph, as in Figure 1.1.1, (perhaps reflected about the t axis) looks like

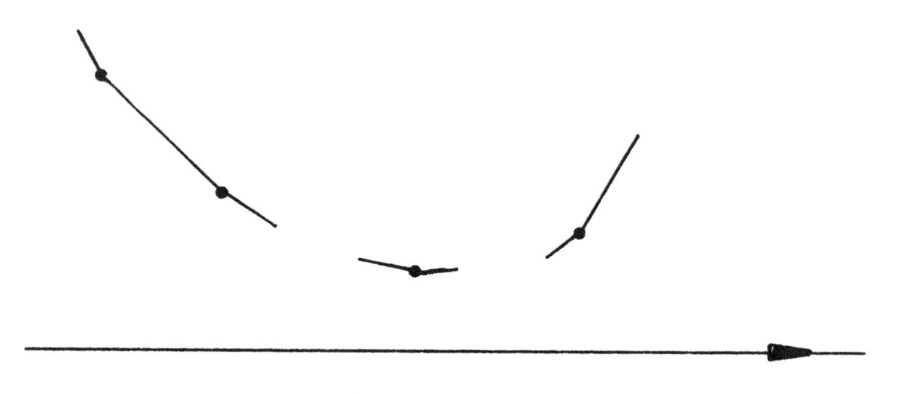

Figure 1 The multiple pivot line search

The BR multiple pivot sequence is $t_i = w_{j+i}/z_{j+i}$, $i = 1,...,m$, ending with the weighted median.

The original BR implementation did not explicitly utilize the fact that t_m is the weighted median of w_i/z_i. It found t_1, the minimum of positive w_i/z_i and tested whether f increased for $t > t_1$. If not, the next smallest positive w_i/z_i was computed, etc. This approach is potentially of quadratic complexity in $(n-k)/2$, the expected number of positive ratios. Simply sorting the ratios reduces this cost to $((n-2) \log (n-2))/2$ and an efficient computation of the weighted median offers scope for further reduction in the required work.

Statement [7] in the algorithm description appears with this flexibility in mind. In fact Armstrong, Frome and Kung (1979) implement BR with a partial sort routine to obtain the weighted median in [7] in an explicit manner. They report timing data consistent with saving $(\log n)/n$ in this part of the computation.

The data structure used in describing the algorithm would not be good in an actual implementation. One undesirable feature is the matrix D_j of edge directions in (11) which is the inverse of $X(\underline{\sigma})$, the matrix made from the rows of X, chosen according to the elements of $\underline{\sigma}$.

The pivot step which produces D_{j+i} from D_j is likely to be numerically unstable and, after many repetitions, may introduce significant roundoff error. A revised simplex approach which would maintain D_j as the product of elementary matrices, or as a product L_jU_j of lower and upper triangular factors, would probably be better. The Armstrong, Frome, and

Kung (1979) implementation of BR incorporates this latter feature.

7.2.2 The Bartels-Conn-Sinclair Algorithm

BR paved the way for treating (1) as an LP, as in (12), but economizing on the straightforward application of the simplex method by using a streamlined data structure and multiple pivot sequences. BCS varies from the standard simplex approach in still another way. It uses a projected gradient idea to choose the direction of descent $\underline{\delta}$ from the current fit $\underline{c}$, to the next fit $\underline{d}$. Another important feature is the attention paid to implementation details. This gives BCS a numerical stability not necessarily present in straightforward implementations of the other methods, and has allowed efficient computations for the manipulations performed. A final property, not described here, is the ability to incorporate extra linear inequalities, thus obtaining linearly constrained LAD fits.

Suppose the current fit is $\underline{c}$, that $\mathbf{Z} = \{i: r_i(\underline{c}) = 0\}$, that $B = \{\underline{x}_i, i \in Z\}$, and that

$$B^{\perp} = \{\underline{u} \in R^k: \langle \underline{u}, x_i \rangle = 0, \text{ all } i \in \mathbf{Z}\}$$

is the orthogonal complement of the basis vectors at $\underline{c}$. In general, if A is a collection of vectors we write

$$A^{\perp} = \{\underline{u}: <\underline{u},\underline{x}> = 0, \text{ all } \underline{x} \in A\}.$$

Let

$$(13) \qquad \underline{h} = -\sum_{i \notin Z} [\text{sign } r_i(\underline{c})]\underline{x}_i$$

From the expression for directional derivatives in (1.3.18), if $\underline{\delta} \in B^{\perp}$

$$(14) \qquad f'(\underline{c},\underline{\delta}) = <\underline{\delta},\underline{h}>.$$

The BCS algorithm chooses $\underline{\delta} = -\mathbf{P}\underline{h}$, the orthogonal projection of $-\underline{h}$ onto $B^{\perp}$. This $\underline{\delta}$ is a <u>projected gradient</u>. If $\underline{\delta} \neq 0$, (14) gives $f'(\underline{c},\underline{\delta}) = -<\underline{\delta},\underline{\delta}> < 0$ so f decreases along $\underline{c} + t\underline{\delta}$, $t > 0$.

The algorithm will choose the next approximation $\underline{d}$ as the minimizer of f on $\underline{c} + t\underline{\delta}$. Because $\underline{\delta} \in B^{\perp}$, $<\underline{d},\underline{x}_i> = y_i$ for $i \in \mathbf{Z}$. In addition the optimal value of t is the weighted median of $\{(y_i - <\underline{c},\underline{x}_i>)/<\underline{\delta},\underline{x}_i>, i \notin \mathbf{Z}\}$ with weights $|<\underline{\delta},\underline{x}_i>|$. If, say, $t = t_q = (y_q - <\underline{c},\underline{x}_q>)/<\underline{\delta},\underline{x}_q>$, $q \notin \mathbf{Z}$, this forces

$$r_q(\underline{d}) = y_q - <\underline{c},\underline{x}_q> - t_q <\underline{\delta},\underline{x}_q> = 0$$

so $\mathbf{Z}$ becomes $\mathbf{Z}\cup\{q\}$ and an additional zero residual has been created at $\underline{d}$.

Returning to (14) suppose the projected gradient $-P\underline{h} = \underline{0}$. This means $\underline{h}$ is in the space spanned by $B = \{\underline{x}_i, i\in\mathbf{Z}\}$, or $\underline{h} = \sum_{i\in Z} \lambda_i\underline{x}_i$, a representation that is unique if $\underline{c}$ is a non-degenerate point: $\dim\{\underline{x}_i, i\in\mathbf{Z}\} = |\mathbf{Z}|$.

For $\underline{\delta} \neq \underline{0} \in R^k$, the directional derivative is, by (1.3.18)

$$f'(\underline{c},\underline{\delta}) = \sum_{i\in Z} |\langle\underline{\delta},\underline{x}_i\rangle| + \langle\underline{h},\underline{\delta}\rangle$$

(15)

$$= \sum_{i\in Z} |\langle\underline{\delta},\underline{x}_i\rangle| + \sum_{i\in Z} \lambda_i \langle\underline{\delta},\underline{x}_i\rangle$$

If $|\lambda_i| \leq 1$, $f'(\underline{c},\underline{\delta}) \geq 0$ for all $\underline{\delta}$, so by Theorem 1.3.6. $\underline{c} \in \mathbf{M}$. If $|\lambda_i| < 1$, $\underline{c}$ is unique.

On the other hand, suppose $|\lambda_j| > 1$. Then let $\underline{\delta}_j = -\text{sign}(\lambda_j)\, P\underline{x}_j$, $P\underline{x}_j$ denoting the orthogonal projection of $\underline{x}_j$ onto

$$\{\underline{x}_i, i \in Z\}, i \neq j\}^{\perp}.$$

By (15) $f'(\underline{c},\underline{\delta}_j) = (1-|\lambda_j|)\langle\underline{\delta}_j,\underline{\delta}_j\rangle$ is negative and the projected gradient $\underline{\delta}_j$ is down hill. If $|\lambda_j| > 1$ for several values of $j \in \mathbf{Z}$, the one with the most negative value of $f'(\underline{c},\underline{\delta}_j)$ is chosen, say $\underline{\delta}_p$.

As before, the next approximation $\underline{d}$ will be a minimizer of f on $\underline{c} + t\underline{\delta}_p$. It will occur at the weighted median $t = t_q = (y_q - <\underline{c},\underline{x}_q>)/<\underline{\delta}_p,\underline{x}_q>$, $q \notin \mathbf{Z}$. This implies $t_q \neq 0$ and $p \notin \mathbf{Z}'$. In fact $\mathbf{Z}$ becomes $(\mathbf{Z}\cup\{q\})\setminus\{p\}$ and q has replaced p in the basic set.

BCS calls a point $\underline{c}$ a dead point in the case we have just considered, where the projected gradient $-\mathbf{P}\underline{h} = \underline{0}$. If $\dim(\underline{x}_i, i \in \mathbf{Z}) < k$ the next approximation $\underline{d}$ may or may not be a dead point. If not, the move along the projected gradient from $\underline{d}$ will fit at least one extra data point. Thus, the normal operation of BCS will produce a non-degenerate extreme point in $j \geq k$ steps. Since $\{\underline{x}_i, i \in \mathbf{Z}\}$ spans R^k for such a point, it is also a dead point. The projected gradient $\underline{\delta}_p$ chosen by BCS in this case is that direction $\underline{\delta}_p$: $<\underline{\delta}_p,\underline{x}_i> = 0$, all $i \neq p$ along which $f'(\underline{c},\underline{\delta}_p)$ is most negative. If the BR algorithm were in Phase II at $\underline{c}$ it would choose precisely the same steepest descent edge and move to the same weighted median t_q on the line $\underline{c} + t\underline{\delta}_p$. Therefore we enunciate the following

> **Observation:** Suppose after $j \geq k$ steps BR is at a non-degenerate extreme point $\underline{c}$. The next iteration of BR agrees with the one BCS would make at $\underline{c}$.

In the normal operation of BR it produces a non-degenerate extreme point $\underline{c}_k$ after k steps and then passes through a sequence $\underline{c}_{k+1},\dots,\underline{c}_N$ of non-degenerate extreme points to $\underline{c}_N \in \mathbf{M}$. If BCS were started at an element $\underline{c}_j$, $j \geq k$ of this sequence, it would generate the remainder of the BR approximations. This shows that although BR and BCS are truly different variants of the simplex method, there is still a rather striking similarity between them. The descent directions at $\underline{c} \in R^k$ can differ in the two algorithms only when BR is in Phase I or if it is in Phase II, at a non-extreme dead point.

The following is a succinct description of the operations performed in the BCS algorithm. So as to not encumber this description, we leave aside for the moment implementation aspects. For example, it is crucial to have a data structure from which the desired quantities may be conveniently extracted at each step. Some of these issues will be addressed subsequently.

Input and Initialization:

[1] Accept $\underline{c} \in R^k$ and $(\underline{x}_i, y_i) \in R^{k+1}$, $i = 1,\dots,n$

[2] $\underline{r} \leftarrow \underline{y} - X\underline{c}$; $Z \leftarrow \{i: r_i = 0\}$

Next Descent Direction:

[3] $\underline{h} \leftarrow - \sum_{i \notin Z} \text{sign } r_i(\underline{c})\ \underline{x}_i$

[4] $\underline{\delta} \leftarrow P\underline{h}$, projection onto $\{\underline{x}_i,\ i \in \mathbf{Z}\}^{\perp}$;
if $\underline{\delta} \neq \underline{0}$ set $p \leftarrow 0$ and go to 8.

[5] Compute $\underline{\lambda} : \underline{h} = \sum_{i \in Z} \lambda_i \underline{x}_i$

[6] $u = (\max |\lambda_i|) = |\lambda_p|$; if $u \leq 1$ go to 11

[7] $\underline{\delta} \leftarrow P\underline{x}_p$, projection onto $\{\underline{x}_i,\ i \in \mathbf{Z},\ i \neq p\}^{\perp}$.

New Basic Vector:

[8] $t \leftarrow (y_q - \langle \underline{c}, \underline{x}_q \rangle) / \langle \underline{\delta}, \underline{x}_q \rangle$, weighted median
of $\{(y_i - \langle \underline{c}, \underline{x}_i \rangle) / \langle \underline{\delta}, \underline{x}_i \rangle,\ i \notin \mathbf{Z}\}$
with weights $\langle \underline{\delta}, \underline{x}_i \rangle$.

Update:

[9] $Z \leftarrow (Z \cup \{q\}) \setminus \{p\}$

[10] $\underline{c} \leftarrow \underline{c} + t\underline{\delta}$; go to 3

Output:

[11] output $\underline{c}$

Some delicacy is required in implementing the foregoing steps. Suppose $\underline{c}_j$ is a non-degenerate extreme point and that $\underline{\sigma} = (\sigma_1,\ldots,\sigma_k)$ contains the elements of the basic set, so $y_{\sigma_i} = \langle \underline{c}_j, \underline{x}_{\sigma_i} \rangle$. The matrix we used to describe the BR algorithm,

$$A_j = \begin{pmatrix} XD_j & \underline{r} \\ \hline D_j & -\underline{c}_j \end{pmatrix},$$

contains all the information required to make the step from $\underline{c}_j$ to $\underline{c}_{j+1}$. Here, writing $X(\underline{\sigma})$ for the matrix with rows $\underline{x}_{\sigma_1},\ldots,\underline{x}_{\sigma_k}$, respectively, we have $X(\underline{\sigma})D_j = I$ and the mth column $\underline{\delta}_m$ of D_j is the descent direction if $\underline{x}_{\sigma_m}$ were to leave the basis. Furthermore the mth column of $X(\underline{\sigma})D_j$ contains the inner products $\langle \underline{x}_i, \underline{\delta}_m \rangle$.

So $f'(\underline{c}, \underline{\delta}_m)$ is readily computed from step [4] in BR and the steepest edge, $\underline{\delta}_p$ selected. Also the inputs to the weighted median calculation are the pth and $k+1^{st}$ columns. Finally, to prepare for the next step, if $\underline{x}_q$ is to enter the basis, the steps [8]-[11] of BR pivot on A_j to produce A_{j+1}.

If $\underline{c}_{j+1}$ is extreme, A_{j+1} will contain all the relevant information to continue from $\underline{c}_{j+1}$. If not, the descent direction $\underline{\delta}$ will be a projector as in step [7] of BCS, and not necessarily a column of D_{j+1}. The required inner products $\langle \underline{x}_i, \underline{\delta} \rangle$ will not necessarily be contained in XD_{j+1} either. Finally at an approximation $\underline{c}$ which is not a dead point, BCS requires data on projectors and their inner products which is not directly available in A. Although these quantities could be explicitly computed from X and information contained in A, a prohibitive amount of extra work might be introduced into such steps of the algorithm.

Projectors may be obtained as follows. The basis is represented in the matrix

$$B = (\underline{x}_{\sigma_1}, \ldots, \underline{x}_{\sigma_m}),\ m \le k$$

whose columns are the points with zero residuals at the current fit. The matrix B is maintained in a factored form as the product

$$(16) \qquad (Q_1 | Q_2) \begin{pmatrix} R \\ O \end{pmatrix}$$

where Q_1 is k by m, Q_2, k by k−m, R, m by m, and O, a k−m by m matrix of zeros. The columns of $(Q_1 | Q_2)$ are mutually orthogonal and R is upper triangular.

With this information at hand, the projector of $\underline{h}$ onto $\{\underline{x}_i,\ i \in Z\}^{\perp}$ is

$$(17) \qquad Q_2 Q_2^T$$

and the λ_i of step [5] of BCS may be obtained by

$$(18) \qquad \underline{\lambda} = R^{-1} Q_1^T \underline{h},$$

'T' denoting transpose.

In the next step a new $\underline{x}_i$ column will either be attached to B or else will replace a column now in B. In both cases, the representation in (16) must be updated to incorporate these changes.

The details of the update are fairly complicated, the aims being numerical stability and then, ease of computation. The actual method used in BCS in fact adopts a representation different from, though related to, that in (16). To carry the description further one would need a familiarity with numerical linear algebra possessed by specialists. For this reason we now refer the interested reader to the BCS reference for the computational details that are recommended in carrying out the algorithm. The description we have given is still useful for highlighting the control steps taken by the algorithm, divorced from any details about how the required computations might be carried out.

The final remark about BCS concerns Step [8], the weighted median calculation. Analysis of their Fortran code reveals that they use a highly specialized technique. The ratios u_i/z_i are maintained in a heap structure [see Knuth (1975)]. The leading item is removed and tested as the weighted median. If the test fails, the heap is updated and the process repeated. One might expect $(n-k)\log(n-k)/2$ comparisons required, on the average, to obtain the weighted median.

7.2.3 The Bloomfield-Steiger Algorithm

The Bloomfield-Steiger algorithm is identical to BR except for two features. It examines the same candidate descent directions as BR. However the heuristic for choosing one of these directions is not "steepest edge", but rather a "normalized steepest edge" criterion. This normalization implies a criterion that is closely related to the idea of weighted medians. This may account for the apparent efficacy of the algorithm. A second difference from BR occurs in its startup phase. BS is not required to increase the number of non-zero components in the next fit if it is judged advantageous to move to a better fit with the same set of non-zero components.

We describe this algorithm using a framework similar to that used in discussing BR. Suppose the current fit is $\underline{c}$, and that $\underline{\delta}_1,\ldots,\underline{\delta}_k$ is a set of directions along which the next iteration could move. The optimum descent direction is that $\underline{\delta}_p$ along which

$$\min\ (f(\underline{c} + t\underline{\delta}_p),\ t \in R) = \min\ [\min\ (f(\underline{c} + t\underline{\delta}_i),\ i \leq k],$$

the inner minimization over t in R. To find it, k weighted median calculations would need to be done, one for each i in the right hand side of the above equation. This costly computation is usually avoided by choosing $\underline{\delta}$ heuristically. The cheaply obtained heuristic choice can be nearly as good as the expensive, optimum one.

The BR heuristic, for example, computes $f'(\underline{c},\underline{\delta}_i)$ for each candidate and selects that $\underline{\delta}_p$ for which f' is least. The whole process is cheap, requiring only $k(n-k)$ additions, as in Step [4] of BR.

To motivate the BS heuristic, suppose $\underline{\delta}_p$ is chosen. Then the line search along $\underline{c} + t\underline{\delta}_p$ seeks the minimizer of

$$f(\underline{c} + t\underline{\delta}_i) = \sum_{i \notin Z} |y_i - \langle \underline{c},\underline{x}_i \rangle - t\langle \underline{\delta}_p,\underline{x}_i \rangle| \tag{19}$$

$$= \sum |u_i - tz_i|$$

It occurs at $t=t^*$, the weighted median of $\{u_i/z_i,\ i \notin Z\}$ with weights $|z_i|$, and $t = 0$ is the current fit.

The weighted median t^* is the median of the distribution function

$$F(t) = \sum_A |z_i| \ / \ \sum_{i=1}^{n} |z_i|, \tag{20}$$

where $\mathbf{A} = \{i: u_i/z_i \leq t\}$. The function F is right continuous and, by definition, $F(t^*) \geq 1/2$ and $F(t) < 1/2$ for $t < t^*$.

BS uses F to assess the potential for decreasing f on $\underline{c} + t\underline{\delta}_p$. The quantity

(21) $$|1/2 - F(0)|,$$

roughly equal to $F(t^*) - F(0)$, is a measure of how far it is possible to move along $\underline{c} + t\underline{\delta}_p$ from the current fit, $t = 0$, to the optimum, at $t = t^*$, expressed in terms of F. The value in (21) therefore may be used as a measure of the relative merit of using $\underline{\delta}_p$ as the descent direction.

If $F(0) = 1/2$ in (21), 0 is already the weighted median for the line search on $\underline{c} + t\underline{\delta}_p$. In non-degenerate situations, $\underline{\delta}_p$ could not be used as a descent direction as it offers no possible improvement from $\underline{c}$. Excluding this case either $F(0) < 0$ or $F(t) \longrightarrow u > 1/2$ as $t \downarrow 0$, by the right continuity of F. The criterion (21) now becomes

(22) $$\begin{array}{ll} 1/2 - F(0) & \text{if } F(0) < 1/2 \\ F(0^-) - 1/2 & \text{if } F(0^-) > 1/2. \end{array}$$

Both expressions are related to directional derivatives at $\underline{c}$. Let $\mathbf{P} = \{i: u_i/z_i > 0\}$ and $\mathbf{N} = \{i: u_i/z_i < 0\}$. Using (19), (20), and the fact that $\mathbf{P} \cup \mathbf{N} \cup \mathbf{Z} = \{1,\dots,n\}$, $1/2 - F(0)$ is

$$1/2 - (\sum_Z |z_i| + \sum_N |z_i|)/ \sum_{i=1}^{n} |z_i|$$

(23) $$= (\sum_P |z_i| - \sum_N |z_i| - \sum_Z |z_i|)/(2 \sum_{i=1}^{n} |z_i|)$$

$$= (\sum_{Z'} z_i \ \mathrm{sign}(u_i) - \sum_Z |z_i|)/(2 \sum_{i=1}^{n} |z_i|)$$

Similar manipulations establish that $F(0^-) - 1/2$ is

(24) $$-(\sum_{Z'} z_i \ \mathrm{sign}(u_i) + \sum_Z |z_i|)/(2 \sum |z_i|)$$

The numerator of the middle line in (23) is minus the right hand derivative of f on $\underline{c} + t\underline{\delta}_p$ (positive for downhill) by (1.3.12a), while the numerator in (24) is the left-hand derivative (also positive for downhill), by (1.3.12b).

Anologous to Step [4] of BR, we compute

$$g_m = \sum_Z |<\underline{\delta}_m, \underline{x}_i>|$$

(25) $$h_m = \sum_{Z'} <\underline{\delta}_m, \underline{x}_i> \ \mathrm{sign}\ (r_i(\underline{c}))$$

$$f_m = \max(h_m - g_m, -h_m - g_m)/ (2 \sum_{i=1}^{n} |<\underline{\delta}_m, \underline{x}_i>|$$

and choose $\underline{\delta}_p$ if $f_p = \max(f_i) > 0$. The BS heuristic for choosing a descent direction is thus based on the "normalized steepest edge" criterion in (25). It has the desirable property of removing any scale differences that exist between the columns of X. It also seems "matched" to the objective of minimizing the LAD distance because it is based on weighted medians.

If the data structure

$$A_j = \begin{pmatrix} XD_j & r_j \\ D_j & -\underline{c}_j \end{pmatrix}$$

is used to represent the information about the fit at step j, the columns of XD_j contain the $<\underline{\delta}_m, \underline{x}_i>$, so (25) is evaluated as easily as the conventional simplex criterion of the BR algorithm. A succinct description of BS, using the notation of BR, is

Input and Initialize:

[1] Accept $(\underline{x}_i, y_i)$, $i = 1,\ldots, n$.

[2] $A \leftarrow \begin{pmatrix} X & \underline{y} \\ I & \underline{0} \end{pmatrix}$

[3] $j \leftarrow 0$, $\underline{\sigma} \leftarrow \underline{0} \in R^k$

Next Descent Direction:

[4] $Z \leftarrow \{i,\ 1 \le i \le n:\ a_{i,k+1} = 0\}$

[5] For m = 1 thru k DO

$g_m = -\sum_Z |a_{i,m}|$

$h_m = \sum_{Z'} a_{i,m} \text{ sign } (a_{i,k+1})$

$f_m = \max (g_m - h_m,\ g_m + h_m)/\sum_{i=1}^{n} |a_{i,m}|$

END

[6] $f_p \leftarrow \max f_i;$ if $f_p \le 0$ go to 10

Next Basic Vector:

[7] $t \leftarrow a_{q,k+1}/a_{qp}$, the weighted median of $\{a_{i,k+1}/a_{ip},\ i = 1,\ldots,n\}$ with weights $|a_{i,k+1}|$

Update:

[8] Pivot (q,p): col p $\leftarrow$ col p/a_{qp}

For $i \ne p,\ 1 \le i \le k+1,$

col i $\leftarrow$ col i $- a_{pi}$ col p

[9] $j \leftarrow j+1$; $\sigma_p \leftarrow q$, go to 4

Output:

[10] $c_i \leftarrow -a_{n+i,k+1}$, $i = 1,\ldots,k$.

The startup phase is finished when no $\sigma_i = 0$. The algorithm operates in the same way, during and after the startup.

While a 'convenient descriptive device,

$$A_j = \left(\begin{array}{cc} XD_j & \underline{r} \\ \hline D_j & -\underline{c}_j \end{array} \right)$$

is likely to be a numerically unstable data structure. Neither LU nor QR factorizations of D_j^{-1} have been attempted for this algorithm. The success of a factored representation in Armstrong Frome, and Kung (1979), and the similarity between the operations performed in BR and BS, suggest that such an implementation might be a worthwhile endeavor.

7.2.4 Summary

In this short section we briefly summarize the main differences between the BR, BCS, and BS algorithms presented in the last section. These may help explain observed differences in performance.

BR starts from $\underline{c}_0 = \underline{0}$. At each step it moves down the steepest edge, with the provision that $\underline{c}_{j+1}$ has at least one non-zero component that was zero in $\underline{c}_j$. BCS can start at any point $\underline{c}_0 \in R^k$. It always moves down a projected gradient. BS starts at $\underline{c}_0 = \underline{0}$ and moves down the normalized steepest edge. This edge need not be restricted to be outside the span of the coordinate vectors corresponding to non-zero components of the current fit.

Typically all three would eventually reach some non-degenerate extreme point. Here BR and BCS both descend down the steepest edge while BS chooses a normalized steepest edge based upon the weighted median heuristic.

The optimum along the descent direction is a weighted median of ratios. BR inefficiently finds the smallest ratio and tests whether it is the optimum. If not, the smallest remaining ratio is found and tested, etc. BCS uses a partial heapsort to explicitly compute the optimum while BS utilizes a weighted median partial quicksort.

These remarks may be summarized in the following table

TABLE 1

	Start	Descent Heuristic		Weighted Median Method
		Startup	at Extreme pt	
BR	$\underline{0} \in R^k$	steepest 'new' edge	steepest edge	successive minima of untested ratios.
BCS	any	projected gradient	steepest edge (equals projected gradient)	partial heapsort
BS	$\underline{0} \in R^k$	normalized steepest edge		partial quicksort

7.3 The Three Algorithms Compared

In this section we describe a study of Anderson and Steiger (1982) which empirically compared BR, BCS, and BS on a variety of LAD curve fits. The results on execution times support surprisingly precise statements concerning the computational complexity of these algorithms.

The published FORTRAN program [Barrodale and Roberts (1974)) was used for BR. Fortran programs of BCS and BS were supplied by the authors. Computations were carried out on a DEC-2060 running under TOPS-20. The reported CPU timings use a subroutine that measures only actual execution time and are not influenced by aspects of the timesharing environment, such as system load.

The three algorithms were compared over a variety of curve fitting problems. Some of these were deterministic and some were random, in which the coefficients of a regression model were estimated from a random sample.

In the deterministic cases the problem is to approximate a function h by $P_k \equiv \sum_{j=1}^{k} c_j \phi_j$, a linear combination of k basis functions ϕ_j. The c_j's are chosen to minimize.

$$(1) \qquad \sum_{i=1}^{n} |h(t_i) - P_k(t_i)|,$$

the t_i being n distinct points. Writing $y_i = f(t_i)$ and $x_{ij} = \phi_j(t_i)$, (1) is the standard LAD distance function

$$(2) \qquad f(\underline{c}) = \sum_{i=1}^{n} |y_i - \langle \underline{c}, \underline{x}_i \rangle|.$$

In the first set of examples, the ϕ_j form the monomial basis: $\phi_j(t) = t^{j-1}$ and P_k is a polynomial of degree k-1. First $h(x) = e^x$ was approximated at n+1 evenly spaced points in

[0,2]. Thus $t_j = 2j/n$, $j=0,\ldots,n$. Table 1 gives the CPU times (net of generating y_i and x_{ij}) and iteration counts for each algorithm over a variety of problem sizes. The number of points, n + 1, ranges as n = 50, 100, 150, 200, 400 and 600. The degree of P_k, k−1, ranges as k = 3, 4, 5, 6, 7, 8.

Table 1

Comparison of CPU Times and Iteration Counts for LAD Approximations of e^x on [0,2] by $\sum c_j t^{j-1}$ Based on n + 1 points $t_i = 2i/n$, $i = 0,1,\ldots,n$.

		3		4		5		6		7		8	
		CPU	ITER	CPU	ITER	CPU	ITER	CPU	ITER	CPU	ITER	CPU	ITER
	BR	.043	5	.062	7	.100	10	.149	14	.181	17	.211	16
50	BCS	.121	7	.238	13	.232	11	.356	15	.407	16	.458	17
	BS	.037	6	.080	11	.107	12	.136	13	.157	13	.229	17
	BR	.109	6	.149	8	.221	9	.331	13	.406	17	.494	17
100	BCS	.223	7	.427	13	.421	11	.570	13	.606	13	.787	16
	BS	.071	6	.163	12	.188	11	.296	15	.339	15	.458	18
	BR	.188	6	.236	7	.398	10	.616	15	.650	16	.905	20
150	BCS	.328	7	.620	13	.696	13	.818	13	.968	15	1.513	22
	BS	.103	6	.239	12	.299	12	.432	15	.502	15	.779	21
	BR	.281	6	.351	7	.598	10	.885	14	.986	17	1.339	20
200	BCS	.432	7	.813	13	.958	14	1.183	15	1.141	13	2.134	26
	BS	.138	6	.317	12	.392	12	.703	19	.699	16	1.021	21
	BR	.845	7	1.083	9	1.032	12	2.688	16	2.656	17	3.755	22
400	BCS	.850	7	1.586	13	1.854	14	2.431	16	2.456	15	3.870	24
	BS	.307	7	.673	13	.888	14	1.476	20	1.484	17	2.139	22
	BR	1.643	7	2.063	9	3.509	11	5.242	16	5.267	20	7.311	23
600	BCS	1.271	7	1.857	10	2.772	14	3.279	14	3.822	16	5.923	25
	BS	.458	7	1.069	14	1.254	13	2.198	20	2.207	17	3.063	21

Table 2 contains the results for the next example. Here

$h(x) = \sqrt{x}$ on $[0,1]$ and $t_j = j/n$, $j = 0,1,\dots,n$. The ranges of n and k are as in Table 1.

Table 2

Comparison of CPU Times and Iteration Counts for LAD Approximation of $\sqrt{x}$ on $[0,1]$ by $\sum c_j t^{j-1}$
Based on n + 1 points $t_i = i/n$, $i = 0,1,\dots,n$.

n		3		4		5		6		7		8	
		CPU	ITER	CPU	ITER	CPU	ITER	CPU	ITER	CPU	ITER	CPU	ITER
50	BR	.067	6	.102	10	.140	13	.167	14	.176	13	.228	16
	BCS	.098	5	.197	10	.237	11	.284	12	.357	14	.417	15
	BS	.038	6	.075	10	.077	8	.129	12	.169	14	.205	15
100	BR	.142	5	.238	10	.280	10	.425	16	.480	15	.541	16
	BCS	.205	6	.412	12	.463	12	.509	12	.736	17	.811	17
	BS	.072	6	.131	9	.175	10	.246	12	.322	14	.434	17
150	BR	.300	6	.445	10	.580	13	.767	17	.836	17	1.001	19
	BCS	.299	6	.557	11	.667	12	.829	14	1.107	18	1.315	20
	BS	.107	6	.193	9	.259	10	.356	13	.447	13	.703	19
200	BR	.413	5	.651	10	.744	9	1.175	18	1.285	15	1.488	18
	BCS	.392	6	.777	12	.880	12	1.195	16	1.504	19	1.578	18
	BS	.141	6	.255	9	.342	10	.577	15	.590	13	.845	17
400	BR	1.607	7	2.158	10	2.738	13	3.387	17	2.633	16	4.435	21
	BCS	.771	6	1.515	12	1.916	14	2.212	15	2.693	17	3.295	20
	BS	.278	6	.546	10	.729	11	1.138	15	1.240	14	1.834	19
600	BR	2.806	5	4.316	12	4.823	10	6.825	18	7.877	17	9.023	20
	BCS	1.154	6	2.258	12	2.868	14	3.287	15	4.339	19	4.888	20
	BS	.420	6	.815	10	1.088	11	1.699	15	2.200	17	2.867	20

In these two examples the columns of the matrix $X = (x_{ij}) \equiv (\phi_j(t_i))$ are nearly linearly dependent. This allows the possibility that round-off errors could become important in the execution of some of the algorithms and thereby seemingly affect their efficiency. Thus, in the third set of examples the

ϕ_j are taken to be the Legendre polynomials shifted onto [0,2] and, as in Table 1, $h(x) = e^x$. The ranges of n and k are as before. BR and BCS had to work harder to produce the same approximations in the (orthonomal) Legendre basis than in the monomial basis, suggesting that conditioning in the original data matrix can indeed affect performance.

Table 3

Comparison of CPU Times and Iteration Counts for LAD Approximation of e^x on [0,2] by $\sum c_j L_j(t)$
Based on n + 1 points $t_i = 2i/n$, $i=0,1,\ldots,n$, L_j the jth Shifted Legendre Polynomial

		3		4		5		6		7		8	
		CPU	ITER	CPU	ITER	CPU	ITER	CPU	ITER	CPU	ITER	CPU	ITER
50	BR	.063	6	.095	10	.124	11	.180	15	.177	13	.308	25
	BCS	.118	6	.205	10	.225	10	.295	12	.289	11	.498	18
	BS	.037	6	.079	11	.107	12	.136	13	.157	13	.229	17
100	BR	.158	6	.222	10	.291	11	.411	14	.426	14	.537	17
	BCS	.229	6	.377	10	.416	10	.572	13	.661	14	.769	15
	BS	.071	6	.163	12	.188	11	.295	15	.339	15	.501	20
150	BR	.286	6	.389	10	.550	13	.717	14	.764	16	1.031	20
	BCS	.339	6	.587	11	.595	10	1.019	17	1.095	16	1.431	20
	BS	.102	6	.239	12	.299	12	.431	15	.501	15	.652	17
200	BR	.408	6	.653	10	.764	12	1.028	16	1.150	17	1.405	20
	BCS	.437	6	.830	12	.832	11	1.443	19	1.571	18	1.950	21
	BS	.138	6	.317	12	.391	12	.703	19	.697	16	.970	20
400	BR	1.329	7	2.104	12	2.249	12	3.209	15	3.903	17	4.554	25
	BCS	.951	7	1.410	10	1.633	11	2.802	19	2.895	17	3.552	19
	BS	.306	7	.672	13	.887	14	1.475	20	1.483	17	2.063	21
600	BR	2.725	7	4.263	12	4.457	12	5.953	17	6.419	19	8.531	25
	BCS	1.433	7	2.430	12	2.441	11	4.174	19	4.720	19	7.36	30
	BS	.475	7	1.067	14	1.252	13	2.195	20	2.198	17	3.192	22

Table 4 depicts the approximation of $h(x) = \sqrt{x}$ on $[0,1]$ by linear combinations of shifted Legendre polynomials. Here BR had a much easier (cheaper) time than in the monomial basis case, while BCS and BS had to work slightly harder to obtain the same approximations.

Table 4

Comparison of CPU Times and Iteration Counts for LAD Approximation of x on $[0,1]$ by $\sum c_j L_j(t)$
Based on n + 1 points $t_i = 2i/n$, $i=0,1,\ldots,n$, L_j the jth Shifted Legendre Polynomial

		3		4		5		6		7		8	
		CPU	ITER	CPU	ITER	CPU	ITER	CPU	ITER	CPU	ITER	CPU	ITER
50	BR	.044	4	.085	9	.090	8	.147	14	.178	14	.221	18
	BCS	.094	5	.159	8	.259	12	.238	10	.381	15	.514	19
	BS	.037	6	.080	11	.107	12	.136	13	.157	13	.228	17
100	BR	.114	5	.198	9	.188	6	.360	16	.393	14	.488	18
	BCS	.172	5	.289	8	.493	13	.489	12	.714	17	.860	18
	BS	.070	6	.163	12	.187	11	.295	15	.339	15	.474	19
150	BR	.189	4	.345	9	.385	9	.525	14	.692	15	.782	17
	BCS	.250	5	.492	10	.752	14	.704	12	.965	15	1.294	20
	BS	.103	6	.240	12	.300	12	.431	15	.496	15	.795	22
200	BR	.308	5	.524	9	.578	9	.911	17	.992	14	1.211	19
	BCS	.329	5	.595	9	.934	13	.912	12	1.391	18	1.809	21
	BS	.137	6	.319	12	.391	12	.702	19	.697	16	.970	20
400	BR	.932	5	1.600	10	1.690	9	2.500	17	3.007	16	3.435	19
	BCS	.723	6	1.250	10	1.919	14	1.996	14	2.572	16	3.219	20
	BS	.309	7	.678	13	.891	14	1.461	20	1.462	17	2.360	25
600	BR	1.878	5	3.230	11	3.354	9	4.757	17	5.996	16	6.665	22
	BCS	.953	5	1.730	9	2.851	14	2.965	14	3.962	17	4.593	19
	BS	.452	7	1.067	14	1.243	13	2.167	20	2.269	18	3.792	27

Finally the three algorithms were compared on LAD

curve fits arising from linear regression models. A sample of size n was generated from the model

$$(3) \qquad Y = c_1X_1 + \dots + c_kX_k + U$$

as follows. For each $i = 1,\dots,n$, successive random numbers $x_{i2}, x_{i3},\dots x_{ik}$, u_i were generated. Then, setting $x_{i1} = 1$, $y_i = c_1x_{i1} + \dots + c_kx_{ik} + u_i$ was completed. The coefficient c_i was chosen to be $i-1$ so that the columns of $X = (x_{ij})$ would be scaled differently. Given such a sample of size n from (3) the three algorithms were used to obtain LAD estimates of the c_i. The CPU times and iteration counts were recorded.

The entire process of generation and estimation was performed 25 times for a particular choice of n and k, and for each algorithm the CPU time and iterations (net of generation) were accumulated. In this way, the comparisons did not depend too strongly on a particular small segment of the random number generator.

To investigate possible effects of the distribution of the random variables X_i,U on the efficiency of the algorithms, the Monte-Carlo studies used members of the Pareto family of densities, p_α, where $p_\alpha(t) = \alpha/(1+t)^{1+\alpha}$, $t \geq 0$, $\alpha \geq 0$. If $\alpha > 1$ the expectation exists and equals $\alpha/(\alpha-1)$ and thus the density

(4) $$p^*_a(t) = a/[1+(t - c)]^{1+a}, \; a \geq 0, \; t \geq c = -a/(a-1)$$

is Pareto and is centered at its mean (of zero).

If $a > 2$, p_a has finite variance and as a rule, the bigger the value of a, the tighter around $a/(a-1)$ are the values of a random variable following p_a. The values $a = 1.2$ and $a = 2.2$ were used in the present experients. In the first case p_a has infinite variance and the random numbers are highly spread. In the latter, the variance is finite and the random numbers are more tightly clustered about their mean.

Table 5 contains the results of the experiments for $a = 1.2$. It presents the total CPU time and total iteration counts (over the 25 repetitions) for each algorithm. In it n ranges through 100, 200, 300, 600, 900, while k takes the values 3, 4, 5, 6, 7, 8.

Table 5

Comparison of CPU Times and Iteration Counts Summed Over 25 Independent Repetitions of Fitting the Model (3), X's and U Pareto (α = 1.2) Distributed

		3		4		5		6		7		8	
		CPU	ITER	CPU	ITER	CPU	ITER	CPU	ITER	CPU	ITER	CPU	ITER
100	BR	2.329	133	3.525	199	4.845	258	6.265	307	7.736	362	9.579	421
	BCS	4.802	150	6.315	185	8.473	235	10.634	276	15.618	385	18.165	427
	BS	1.634	130	2.835	189	4.280	243	5.807	287	8.164	359	9.929	388
200	BR	5.810	148	8.712	230	11.406	277	14.757	341	17.677	387	23.150	477
	BCS	8.800	142	13.849	218	17.267	257	23.305	329	29.228	392	35.409	454
	BS	3.448	143	5.993	208	9.250	274	11.957	304	15.599	351	20.912	422
300	BR	10.716	165	15.685	240	21.072	303	25.628	369	30.722	410	35.965	465
	BCS	14.919	169	19.955	215	28.264	289	33.765	325	41.919	383	55.079	487
	BS	5.904	170	9.416	221	13.581	270	18.593	321	24.724	376	31.094	422
600	BR	30.763	181	39.922	226	51.898	296	68.659	392	81.507	454	96.727	537
	BCS	29.864	172	40.061	219	56.184	298	67.741	335	88.994	422	109.205	500
	BS	11.692	170	18.580	220	28.421	287	37.661	328	52.287	404	67.960	470
900	BR	54.228	173	73.797	232	91.825	297	114.026	366	141.273	474	163.161	509
	BCS	42.613	161	60.526	224	81.709	289	104.158	347	137.131	434	164.120	508
	BS	16.898	162	27.601	220	42.545	287	56.133	327	76.461	395	98.191	452

Table 6 contains the results of estimating the regression in (3) when the variables are centered Paretos with α = 2.2, while in Table 7, the variables were unit normal. The ranges of n and k are as in Table 5.

Table 6

Comparison of CPU Times and Iteration Counts Summed Over 25 Independent Repetitions of Fitting the Model (3), X's and U Pareto (α = 2.2) Distributed

		3		4		5		6		7		8	
		CPU	ITER	CPU	ITER	CPU	ITER	CPU	ITER	CPU	ITER	CPU	ITER
100	BR	2.516	166	3.272	191	4.781	270	7.132	392	8.445	418	10.952	510
	BCS	4.861	153	7.154	215	9.872	282	13.159	353	16.055	401	18.235	431
	BS	1.869	156	2.876	194	4.827	282	7.038	357	8.907	396	10.933	432
200	BR	5.495	155	8.727	255	11.678	307	16.454	419	19.283	453	24.793	584
	BCS	9.776	162	15.292	249	18.882	289	29.804	440	33.701	466	40.671	534
	BS	3.729	160	6.744	242	9.128	272	14.626	383	18.139	415	25.448	523
300	BR	10.061	179	14.932	254	19.572	317	25.294	394	31.669	458	43.595	620
	BCS	15.861	184	23.085	256	34.307	370	40.484	404	49.944	471	61.188	556
	BS	5.924	173	10.056	242	15.088	306	22.545	398	29.633	460	39.100	541
600	BR	27.896	191	41.928	302	53.599	366	70.324	463	86.768	550	109.447	691
	BCS	31.951	188	52.694	309	69.280	384	85.736	442	107.685	530	129.927	615
	BS	12.249	182	23.181	287	32.547	336	50.467	454	61.748	486	86.095	606
900	BR	53.620	205	77.816	294	100.714	392	124.573	479	159.659	618	207.675	810
	BCS	53.679	220	76.746	298	94.919	349	140.739	498	174.045	582	199.885	640
	BS	19.799	201	31.478	258	50.909	354	69.596	417	98.602	522	145.888	692

Table 7

Comparison of CPU Times and Iteration Counts Summed Over 25 Independent Repetitions of Fitting the Model (3), X's and U Distributed as Unit Normals

		3		4		5		6		7		8	
		CPU	ITER	CPU	ITER	CPU	ITER	CPU	ITER	CPU	ITER	CPU	ITER
100	BR	2.628	189	4.532	309	6.051	371	7.675	430	10.414	545	13.102	639
	BCS	4.763	156	6.586	203	10.390	309	13.748	379	15.691	401	20.017	487
	BS	2.076	179	3.511	248	5.669	339	8.265	427	11.301	513	15.765	640
200	BR	6.501	220	10.645	341	14.967	431	21.429	585	27.450	700	31.739	747
	BCS	8.873	153	14.128	238	19.490	312	27.199	410	33.629	475	42.192	569
	BS	4.057	180	8.116	300	12.426	384	20.096	541	26.726	630	33.943	711
300	BR	11.397	244	18.377	356	24.702	466	34.960	598	44.996	720	56.087	852
	BCS	13.729	163	22.064	257	30.940	344	41.183	427	52.011	510	61.594	575
	BS	6.365	192	12.276	305	21.034	443	30.295	549	41.563	661	55.173	780
600	BR	29.117	246	51.092	423	71.403	568	94.026	708	115.517	824	148.900	1020
	BCS	28.032	172	44.059	264	65.065	377	89.161	485	110.459	564	137.302	674
	BS	14.256	223	29.182	372	44.597	478	67.771	625	91.962	742	125.849	904
900	BR	56.600	273	92.479	442	130.753	614	164.854	727	202.856	868	258.478	1072
	BCS	46.146	195	65.847	265	100.534	394	129.853	475	153.596	527	204.203	679
	BS	22.570	238	43.286	370	71.677	516	105.188	651	142.546	772	195.548	941

Here are some further details concerning the last three experiments. Random numbers were generated using the FORTRAN function RAN on the DEC 20. In Tables 5, 6, and 7 each cell (n,k) uses a different segment of the generator. But within a cell, each algorithm solved the fitting problems generated by the same random numbers. Random numbers with density p_a^* have distribution function

$$P_a^*(t) = 1 - 1/[1 + (t-c)]^a,$$

Hence, if U is uniform [0,1]

$$U^{-1/a} + c - 1$$

will have centered, Pareto density, (4). This was the generation method used in Tables 5 and 6. Normal random numbers were approximated by sums of 12 independent uniforms, less the constant 6.

Anderson and Steiger studied the data in Tables 1–7 to try to describe the behavior of BR(n,k), BCS(n,k) and BS(n,k), the computational complexity of the algorithms for problems of size (n,k), complexity being measured by CPU time. Examination of the columns of all tables reveals a characteristic difference between BR and the other two algorithms. For all k, as n increases, BR increases faster than linearly (perhaps like n^2), while BCS and BS increase linearly. The algorithms BCS and BR differ only in how the descent direction is chosen at non-extreme fits (startup) and in the weighted median calculation. The startup differences are not likely to be significant enough to explain these performance comparisons. It is therefore probable that an efficient line search – as in the

Armstrong-Frome (1976) modification of BR - would render BR quite comparable to BCS. In fact Armstrong, Frome, and Kung (1979) report savings over BR consistent with this suggestion.

To reinforce the proposal that BCS and BS grow linearly in n, LAD straight lines were fit to (n, BCS(n,k)) and (n, BS(n,k)) for each column k in each of Tables 1-7. The data in Tables 5-7 were first divided by 25 to make the timings comparable to those in Tables 1-4. In all cases, CPU times were extremely well described and departures from the linear fits were negligible

$$\begin{aligned} BCS(n,k) &= a_k(h)n + b_k(h) \\ BS(n,k) &= a_k^*(h)n + b_k^*(h). \end{aligned} \qquad (5)$$

The slopes $a_k(h)$ and $a_k^*(h)$ depend on the number of parameters, k, being fit and on h, a variable that describes the curve-fitting problem (eg. e^x approximated by k^{th} degree polynomials, k^{th} order regression with Pareto (a=1.2) errors, etc.)

Table 8 shows $a_k(h)$ and $a_k^*(h)$ for all the values of k that were considered and in each curve-fitting problem that was studied. In all cases, $a_k^*(h) < a_k(h)$, and the difference is usually sizeable. For example, when k = 3 and one is approximating e^x in the Legendre basis (Table 3), BS grows with n at less than 1/3 the rate that BCS does.

Table 8

Slopes of the Straight-Line Relationships of BCS(n,k) and BS(n,k) with n for various k. (See eq. 5)

Table		3	4	5	6	7	8
1	BCS	.00210	.00382	.00461	.00531	.00643	.00994
	BS	.00077	.00170	.00212	.00375	.00374	.00511
2	BCS	.00191	.00372	.00481	.00546	.00721	.00813
	BS	.00070	.00137	.00184	.00285	.00306	.00487
3	BCS	.00241	.00410	.00404	.00701	.00806	.01319
	BS	.00078	.00170	.00212	.00374	.00372	.00556
4	BCS	.00156	.00284	.00471	.00496	.00651	.00733
	BS	.00076	.00172	.00210	.00369	.00393	.00642
5	BCS	.00189	.00270	.00366	.00468	.00617	.00730
	BS	.00076	.00123	.00191	.00252	.00345	.00442
6	BCS	.00251	.00351	.00425	.00638	.00802	.00910
	BS	.00092	.00143	.00239	.00314	.00460	.00688
7	BCS	.00213	.00296	.00463	.00587	.00686	.00926
	BS	.00106	.00201	.00338	.00486	.00662	.00923

BR(n,k), BCS(n,k) and BS(n,k) all increase with k in a linear fashion, as can be seen by graphing the rows in each table against k. While LAD straight lines do not fit these data quite as well as they did with the plots of CPU time versus n, the fits are still excellent. However, in the deterministic context (Tables 1–4) there is a slight tendency for all the algorithms to work proportionately harder for even k then for odd k. There seems to be no ready explanation for this curious fact.

Dividing the data in Tables 5–7 by 25 so they are comparable to the CPU times for the deterministic context portrayed in Tables 1–4, we see that all the algorithms had to work harder to obtain the regression fits than the deterministic ones. The Pareto (α = 1.2) case was the easiest regression and BR and BS had the most difficulty with the Normal case (Table 7). This may be due to the fact that $(\underline{x}_i, y_i)$ points are highly dispersed in the former case. In general, the k points determining the optimal hyperplane will tend to be "spread out", with high inter–point distances. With Normal data many sets of k points are nearly equal, high dispersion, and define near–optimum fits. But with infinite variance Pareto data, only a few sets of k points are nearly optimal so the task of discriminating is simpler in this case.

In view of (5) and the linearity of all the algorithms in k, Anderson and Steiger tried to describe BCS and BS by

$$
\text{(6)} \qquad \begin{aligned} BCS(n,k) &= a + bn + ck + dnk \\ BS(n,k) &= a^* + b^*n + c^*k + d^*nk. \end{aligned}
$$

In each table these parameters were estimated by minimizing

$$
g(a^*,b^*,c^*,d^*) = \sum_{n,k} |BS(n,k) - (a^* + b^*n + c^*k + d^*nk)|
$$

The models in (6) fit the table data extremely well. To measure how well, an LAD analogue of standard deviation is used. In Table 5 for example, the median CPU time for BCS is 1.273 and

$$
\sum_{n,k} |BCS(n,k) - 1.273| = 34.768
$$

measures the variability of the CPU times for the BCS algorithm over the Pareto (α = 1.2) regression problems. The variability measure of the difference between BCS and the right hand side of (6) is 2.36 so the model in (6) may be said to "explain" 93.21% of the variability of the CPU times. In a similar way the model for BS in (6) explains 91.38% of the variability of the data in Table 5.

The parameter values for the models in (6) are given in

Table 9, along with the percent of CPU time variability that is explained by that model. This is done for each curve-fitting problem summarized in Tables 1–7. The information would account for the linearity of the algorithms in k and in n. Furthermore comparing d and d^*, it suggests that BS grows slowest with size nk.

Table 9

Parameters Fit to the Complexity Model (14) for each curve fitting context.

		a	b	c	d	Percent Explained
1	BCS	.0100	-.0017	.0016	.00125	84.2
	BS	.0350	-.0019	-.0123	.00089	87.8
2	BCS	.0844	-.0019	-.0281	.00142	80.5
	BS	.0415	-.0019	-.0171	.00092	87.4
3	BCS	.0196	-.0019	-.0015	.00126	87.7
	BS	.0395	-.0017	-.0104	.00077	89.4
4	BCS	-.0398	-.0020	.0163	.00121	90.4
	BS	.0471	-.0022	-.0178	.00098	84.2
5	BCS	.0546	-.0019	-.0170	.00115	93.2
	BS	.0442	-.0017	-.0129	.00074	91.4
6	BCS	.0362	-.0019	-.0263	.00135	94.3
	BS	.0928	-.0027	-.0264	.00102	87.1
7	BCS	.1297	-.0031	-.0372	.00153	93.7
	BS	.1635	-.0040	-.0555	.00151	89.9

7.4 Slightly Overdetermined Equation Systems

In seeking the minimizers of

$$(1) \qquad f(\underline{c}) = \sum_{i=1}^{n} |y_i - \langle \underline{c}, \underline{x}_i \rangle|, \quad \underline{c} \in R^k$$

it seems natural that the complexity of the task would increase with k, the number of parameters being fit. In fact the study described in Section 3 asserted that the complexity of the BR, BCS, and BS algorithms all increase linearly in k. If for n points in R^k the computational cost were C, the extra cost with 2j more parameters would be twice the extra cost with j more parameters, for each of the algorithms studied.

In this section we describe a simple idea of Seneta and Steiger (1983) which casts (1) as an equivalent, complementary problem that becomes easier to solve as k increases. Using matrix notation, (1) may be written as the unconstrained problem over R^k

$$(2) \qquad \text{minimize } \|\underline{r}(\underline{c})\|, \quad \underline{c} \in R^k.$$

where $\underline{r}(\underline{c}) = \underline{y} - X\underline{c}$ and $\|\underline{u}\| = \sum_{i=1}^{n} |u_i|$.

If X has $p \leq k$ independent columns and A is an n−p by n matrix whose rows are orthogonal to the span of the columns of X, then AX is an n−p by k zero matrix and $A\underline{r} = A(\underline{y} - X\underline{c}) = A\underline{y}$. Defining $\underline{b} = A\underline{y}$ and supposing that the rows of A are independent, we see that (2) is equivalent to the linearly constrained problem

$$\text{(3)} \quad \begin{aligned} &\text{minimize } \|\underline{r}\|, \quad \underline{r} \in R^n \\ &\text{subject to } A\underline{r} = \underline{b} \end{aligned}$$

The equivalence is trivial because

$$\{\underline{r} \in R^n: \underline{r} = \underline{y} - X\underline{c},\ \underline{c} \in R^k\} = \{\underline{r} \in R^n: A\underline{r} = \underline{b}\}.$$

For if $\underline{r}$ solves (3) $A\underline{r} = \underline{b} = A\underline{y}$ so $\underline{r} - \underline{y}$ is in the p dimensional orthogonal complement of the row space of A, or a linear combination of the columns of X, so $\underline{r} = \underline{y} - X\underline{c}$ for some $\underline{c} \in R^k$. Since $\underline{c}$ is suppressed, (3) might be called the parameter−free, or Cauchy form of the problem. We will call (1) the standard, or Gauss form [see Seneta (1976)].

The equivalence between the two forms of a LAD curve fit provide further information about solutions of (3). By

Theorem 1.3.1, if $\underline{r}(\underline{c})$ solves (2), at least p of its components may be taken to be zero. The same property must hold for solutions to (3).

Algorithms for (1) and (2) move to the optimum via a sequence $\underline{r}_1, \underline{r}_2, \ldots, \underline{r}_N$ of approximations for which $||(\underline{r}_s)|| \geq ||(\underline{r}_{s+1})||$ and at least p components of $\underline{r}_s$ are zero. Accordingly, taking X to be of full rank, we see that if the jth component of $\underline{r}_s$ is zero for $j \in B = \{j_1, \ldots, j_k\}$, the $\underline{c}$ that minimizes f is a solution of the system

$$\sum_{j=1}^{k} c_j \, x_{ij} = y_i, \ i \in B.$$

The non-zero components of the optimal $\underline{r}$ satisfy

$$r_i = y_i - \sum_{j=1}^{k} c_j x_{ij}, \ i \notin B.$$

However if $\underline{c}$ has not been explicitly computed, these r_i's are also the solutions to $\sum_{j \notin B} a_{ij} r_j = b_i$.

The potential advantage in considering (3) rather than (2) is based on the following observation. Writing the LAD problem in parametric form (1) requires the n by (k+1) augmented matrix $(X|\underline{y})$. If instead it is written in Cauchy form (3), the augmented matrix $(A|\underline{b})$ of size n−p by n+1 is required. Assuming p=k, the latter is smaller when $k \geq n/2$.

This suggests that a good algorithm for (3) might outperform the best procedures for (1) as $k \geq n/2$ increases, even taking account of the cost in finding A and $\underline{b}$. The surprise is that the cross-over point seems to be about $k = n/3$.

Here is an algorithm for (3), in all respects similar to the Bloomfield-Steiger algorithm for (2). Suppose $A^* = (A|\underline{b})$ is given, A an m by n > m matrix of full rank, and $\underline{b} \in R^m$.

As an initial solution take

$$\underline{r} = \begin{pmatrix} \underline{z} \\ \\ \underline{0} \end{pmatrix}$$

$\underline{0} \in R^{n-m}$ a zero vector, and write $A = (B|N)$, where B is the n by m matrix consisting of the first m columns of A, assumed without loss of generality to be linearly independent, and N the remaining n−m columns. The columns of B form the initial <u>basis</u> and correspond to the m (generally) non-zero elements of $\underline{r}$. Because $A\underline{r} = \underline{b}$, $\underline{z} = B^{-1}\underline{b}$. The <u>basic set</u> **B** = {1,...,m}, at the start.

To continue, a column of B, say the pth, will leave the basis via an exchange with a column of N, say the qth, that will enter: z_p will become 0 and 0_q will become $t \neq 0$. The choice of q is heuristic and once it is determined, p is optimally chosen.

Assuming that $\underline{n}_q$ has been chosen to enter, the next approximation will be found as a member of the one-parameter family

$$\underline{r}^*(t) = \begin{pmatrix} \underline{z}^*(t) \\ \\ \underline{0} \end{pmatrix} + t\underline{e}_{m+q}, \quad t \in R,$$

where $\underline{e}_i \in R^n$ is the ith unit coordinate vector and $\underline{z}^*(0) = \underline{z} = B^{-1}\underline{b}$. Thus $\underline{r}^*(0)$ is the current solution.

Since $A\underline{r}^*(t) = \underline{b}$, $B\underline{z}^*(t) + t\underline{n}_q$ must equal $\underline{b}$, $\underline{n}_q$ denoting the qth column of N. Thus

$$(4) \qquad \underline{z}^*(t) = B^{-1}\underline{b} - t\, B^{-1}\underline{n}_q.$$

The value of g in (3) will now be

$$g(\underline{r}^*(t)) = \sum_{i=1}^{n} |r_i^*(t)|$$

$$(5) \qquad = \sum_{i=1}^{m} |z_i^*(t)| + |t|$$

$$= \sum_{i=1}^{m} |z_i - tv_i| + |t|$$

where we write $\underline{v} = B^{-1}\underline{n}_q$. This is a line search. The value of t that minimizes (5) defines a line $y = tx$ that is the LAD fit through the origin for the points (v_i, z_i), $i = 1,\ldots,m$ and $(1,0)$. It is easily recognized to be the weighted median of the ratios z_i/v_i, with weight $|v_i|$, and 0, with weight 1.

If the weighted median is zero, $\underline{n}_q$ may not enter the basis so another column of N would be investigated. Otherwise $t = t_p = z_p/v_p$, for some p, $1 \le p \le m$, and the pth term in the sum in (5) becomes zero. This means that $z_p^*(t_p) = 0$ so that $\underline{n}_q$ replaces $\underline{b}_p$ (the pth column of B) to form the next basis. The objective in (5) cannot increase because when $t = 0$, $\underline{r}^*(t) = \underline{r}$ is the current solution and $g(\underline{r}^*(0)) \ge g(\underline{r}^*(t_p))$.

In choosing $\underline{n}_q$, the non-basic column that will enter, one seeks that column which, when it optimally replaces some basic column, will produce the greatest reduction in the objective (3). A brute force "look ahead" method would compute $\underline{v} = B^{-1}\underline{n}_q$ for each column $\underline{n}_q$ of N, minimize (5), and then enter that column corresponding to the smallest value of g. A heuristic method can avoid the computational cost of actually minimizing (5) for each $\underline{n}_q$, and would usually choose the best column anyway.

The following method is based on weighted medians. If $\underline{n}_q$ were chosen, the foregoing procedure for optimally selecting a column to leave the basis would compute $\underline{v} = B^{-1}\underline{n}_q$ and minimize (5). The minimum occurs at t^*, the weighted median of the ratios z_i/v_i with weights $|v_i|$, and 0 with weight 1. Assume it is not equal to zero or else $\underline{n}_q$ cannot enter. It may be expressed as the median of the distribution function

$$(6) \qquad F(t) = (I_0(t) + \sum_A |v_i|)/(1 + \sum_{i=1}^{m} |v_i|) ,$$

where $I_0(t)$ is 1 on $[0,\infty)$ and 0 otherwise, and

$$A = \{i : z_i/v_i \leq t\}.$$

By definition, $F(t^*) \geq 1/2$ and $F(t) < 1/2$ for $t < t^*$.

In (4) the current approximation corresponds to $t = 0$, and the next approximation to t^*. For this reason the quantity

(7) $\quad |1/2 - F(0)|,$

approximately $|F(t^*) - F(0)|$, is a rough measure of the distance – in terms of F – between the current approximation and the next one. It attempts to measure the relative advantage of using $\underline{n}_q$ to optimally replace a basic column.

If in (6) $F(0) = 1/2$, replacement with $\underline{n}_q$ would not decrease g in (5). Excluding this case, either $F(0) < 1/2$ or $F(t) \downarrow u > 1/2$ as $t \uparrow 0$, by the right continuity of F. The criterion in (7) would then become

$$(8) \quad \begin{cases} 1/2 - F(0) & \text{if } F(0) < 1/2 \\ F(0^-) - 1/2 & \text{if } F(0^-) > 1/2 \end{cases}$$

As in subsection 7.2.3 both expressions are normalized gradients. The quantity $1/2 - F(0)$ may be written

$$(9) \quad \Big(\sum_{Z'} v_i \operatorname{sign}(z_i) - 1 - \sum_{Z} |v_i|\Big)/\Big(2\big(1 + \sum_{i=1}^{m} |v_i|\big)\Big)$$

and $F(0^-) - 1/2$, as

$$(10) \qquad -(\sum_{Z'} v_i \operatorname{sign}(z_i) + 1 + \sum_{Z} |v_i|)/(2(1 + \sum_{i=1}^{m} |v_i|))$$

where $\mathbf{Z} = \{i,\ 1 \le i \le m\colon z_i = 0\}$ and $\mathbf{Z}' = \{1,\dots,m\}\setminus\mathbf{Z}$. The numerator in (9) is minus the right hand derivation of (5) at $t = 0$. It measures the rate of decrease in g as one moves away from the current approximation in the direction

$$\underline{w}_q = \underline{r}^*(1) - \underline{r}^*(0) = \begin{pmatrix} \underline{z}-\underline{v} \\ \\ \underline{0} \end{pmatrix} + \underline{e}_{m+q}$$

where $\underline{v} = B^{-1}\underline{n}_q$. Similarly the left hand derivative at $t = 0$ in (5) is the numerator of (10), so it measures the rate of decrease as one moves away from the current approximation in the direction $-\underline{w}_q$.

Again, analogous to equation (25) in subsection 7.2.3, the normalized steepest edge test computes

$$g_j = 1 + \sum_{Z} |v_i^{(j)}|$$

$$(11) \qquad h_j = \sum_{Z'} v_i^{(j)} \operatorname{sign}(z_i)$$

$$f_j = \max(h_j - g_j, -h_j - g_j)/(2(1 + \sum_{i=1}^{m} |v_i^{(j)}|))$$

where $\underline{z} = B^{-1}\underline{b}$ and $\underline{v}^{(j)} = B^{-1}\underline{n}_j$. The heuristic chooses $\underline{n}_q$ to enter if $f_q = \max(f_i, i = 1,\ldots,n-m)$. If $f_q \le 0$ and no component of $\underline{z}$ is zero, $\underline{r}$ is optimal by the equivalence of (2) and (3) and Corollary 1.3.1. Otherwise $\underline{n}_q$ replaces $\underline{b}_p$ and the process continues.

A convenient data structure for describing the operation of the algorithm begins with the augmented matrix $A^* = (B|N|\underline{b})$ premultiplied by B^{-1}. The matrix

$$(12) \qquad D = (I|B^{-1}N|B^{-1}\underline{b})$$

contains the information that the vector $\underline{\sigma} = (\sigma(1),\ldots,\sigma(m))$ of indices pointing to the basic columns is $(1,\ldots,m)$, that the vector $\underline{\sigma}^c = (\sigma^c(1),\ldots,\sigma^c(n-m))$ of indices pointing to non-basic columns is $(m+1,\ldots,n)$, and that $\underline{z} = B^{-1}\underline{b}$ gives the values of the non-zero components of $\underline{r}$. Thus $r_{\sigma(i)} = z_i$ and $r_{\sigma^c(i)} = 0$.

To find the column of N that will become basic, the criterion f_j in (11) is evaluated for each $j = 1,\ldots,n-m$. The $\underline{v}$'s are the columns of $B^{-1}N$, from (4).

Once the qth element of $\underline{\sigma}^c$ (call it j; initially $j = \sigma^c(q) = m+q$) has been selected to enter, and the pth element of $\underline{\sigma}$ is chosen to leave, (12) is updated by a pivot step using Jordan elimination : the pth row of D is multiplied by a constant so the jth column entry is 1; then multiples of row p are added to each other row so that j^{th} column entries become zero.

If E is the m by m matrix that effects these steps, D becomes

$$(13) \qquad D^* = ED = (E|EB^{-1}N|EB^{-1}b).$$

The new basis pointers are $\underline{\sigma}^* = (1,\dots,p-1,m+q,p+1,\dots,m)$ and the non-basic ones are $(\underline{\sigma}^*)^c = (m+1,\dots,m+q-1,p,m+q+1,\dots,n)$. The matrix EB^{-1} is the inverse of the new basis matrix $B^* = A(\underline{\sigma}^*)$, formed by taking columns from A according to $\underline{\sigma}^*$, and $\underline{z}^* = EB^{-1}\underline{b}$ are the non-zero components of the new solution $\underline{r}^*$. In general, after several steps $\underline{\sigma}$ points to the m columns of the matrix that comprise an identity, $\underline{\sigma}^c$ to the n−m columns corresponding to $B^{-1}N$ for the current basis/non-basis partition of A, and the n + 1st column gives the values of the basic residuals, $r_{\sigma(i)}$. The rest, $r_{\sigma^c(i)}$, equal zero.

Suppose we are given X and $\underline{y}$ in (1), X an n by k matrix of full rank. To convert the LAD fitting problem into (3) we require an n−k by n matrix A of full row rank that satisfies AX = 0. Write A = (B|N), B, n−k by n−k, and

$$X = \begin{pmatrix} X_B \\ X_N \end{pmatrix}$$

where X_B is n−k by k and X_N is k by k, and we suppose without loss of generality that X_N is invertible. Then $BX_B + NX_N = 0$ and if we take $B = I$, $NX_N = -X_B$. Transposing, we define N by

$$(14) \qquad X_N^T N^T = -X_B^T ,$$

n−k linear systems each of size k. If we used Gaussian elimination, and backsolved, N could be found in about $k^3/3 + (n-k)k^2 = nk^2 - 2k^3/3$ multiplication and division operations. This initializes (12) with $(I|N|\underline{b})$, where $\underline{b} = A\underline{y}$.

It is interesting that the first extreme solution using BS directly on (1) would cost about nk^2 steps. For each of k iterations, pivots with complexity nk are performed. Therefore translation from (1) to (3) has about the same computational cost as that required for BS to find the first extreme fit.

The foregoing discussion may be summarized more succinctly, as in the following formal description of the algorithm.

Initialize:

[1] Accept $(\underline{x}_i, y_i) \in R^{k+1}$, $i = 1,\dots,n$,

$m \leftarrow n-k$

[2] Renumbering the rows of $(X|\underline{y})$ if necessary so that X_N, the bottom k rows of X, is invertible, find N using (14).

[3] $D \leftarrow (A|\underline{b})$ where $A \leftarrow (I|N)$ is m by n and $\underline{b} = A\underline{y}$.

[4] $\underline{\sigma} \leftarrow (1,\dots,m)$, $\underline{\sigma}^c \leftarrow (m+1,\dots,n)$.

[5] $r_{\sigma(i)} \leftarrow b_i$, $i = 1,\dots,m$, $r_{\sigma^c(i)} \leftarrow 0$, $i = 1,\dots,n-m$

Choose Entering Column:

[6] $Z \leftarrow \{i,\ 1 \le i \le m: D_{i,n+1} = 0\}$

[7] For j = 1 thru k DO

$v_i \leftarrow D_{i,\sigma^c(j)}$, i j= 1,...,m

$g_j = 1 + \sum_Z |v_i|$

$h_j = \sum_{Z'} v_i \text{ sign } (D_{i,n+1})$

$f_j = \max(h_j - g_j, -h_j - g_j)/2(1 + \sum_{i=1} D_{i,n+1})$

END

[8] $f_q = \max(f_j,\ 1 \le j \le k)$

[9] If $f_q > 0$ go to 11

[10] If $(\prod_{i=1}^{m} r_{\sigma(i)}) = 0$ return "degenerate",
else $r_{\sigma(i)} \leftarrow D_{i,n+1}$, $r_{\sigma^c(i)} \leftarrow 0$ and
SOLVE $y_{\sigma^c(i)} = \langle \underline{c}, \underline{x}_{\sigma^c(i)} \rangle$, $i = 1,\ldots,k$
for $\underline{c}$ and STOP.

Find Leaving Column:

[11] $v_i \leftarrow D_{i,\sigma^c(q)}$, $i = 1,\ldots,m$

[12] $\hat{t} \leftarrow$ weighted median of $D_{1,n+1}/v_1,\ldots,D_{m,n+1}/v_m$
and 0 with weights $|v_1|,\ldots,|v_m|$ and 1,
say $\hat{t} = D_{p,n+1}/v_p$

Update:

[13] divide row p of D by $D_{p\sigma^c(q)}$

[14] For $i \ne p$, (row i of D) $\leftarrow$ (row i of D) −
(row p of D)$*D_{i\sigma^c(q)}$

[15] $\sigma(p) \leftrightarrow \sigma^c(q)$

[16] Go to 6

Steps [1]–[4] may be executed more easily by performing Gauss-Jordan elimination on X^T. This produces A with an identity matrix in the columns corresponding to $\underline{\sigma} = (\sigma(1),\dots,\sigma(m))$ and N in the remaining columns. $A\underline{y}$ gives $\underline{b}$ and the rest of the algorithm proceeds as described above.

For numerical stability, $D = B^{-1}(A|\underline{b})$ might better be represented using a factored form of B^{-1}, either as LU or QR. These implementations have not been tried.

The SS algorithm could also be cast as an analogue of Barrodale-Roberts if a "steepest edge" criterion chooses the entering column. In this case Step 6 would compute $f_j = \max(h_j - g_j, -h_j - g_j)$. The current version of SS works a little harder to compute the f_j but this might be repaid in over-all efficiency.

Seneta and Steiger compared the computational cost of BS in solving (1) to that of SS in solving (3). The results are somewhat surprising.

One iteration of BS costs (in terms of multiplications and

divisions) $n(k+1)$ operations to update, plus the cost, $WM(n-k)$ of forming $n-k$ ratios and then finding the weighted median. The first k iterations successively pivot new points into the fit and may be regarded as a start up phase, at total cost of $k[n(k+1)+WM(n-k)]$. If N further iterations are required the total work for BS would be

$$(15) \quad k[(n(k+1)+WM(n-k)]+N[n(k+1)+WM(n-k)]+k^3/3 + k^2.$$

the last two terms reflecting the extra cost of solving $\underline{y} = X\underline{c}$ for $\underline{c}$, once it is known which k rows of X determine the optimal fit.

To initialize SS for (3) it costs $nk^2 - 2k^3/3$ operations to obtain A plus a further $(n-k)k$ to compute $\underline{b} = A\underline{y}$. Thereafter, each iteration then costs $(n+1)(n-k) + WM(n-k)$, the first term being the cost of updating D in (14). Thus if M iterations of SS are required, the total work would be about

$$(16) \quad \begin{aligned} &nk^2 + nk - 2k^3/3 - k^2 + M[(n+1)(n+k) + WM(n-k)] + \\ &(n-k)^3/3 + (n-k)^2 \end{aligned}$$

The difference between the BS and SS for just the start up phase and the final solve is therefore

$$k\ WM(n-k) + 2k^3/3 + k^2 + [k^3 - (n-k)^3]/3 \tag{17}$$

$$(k^2 - (n-k)^2).$$

Since WM(n−k) is n−k divisions for the ratios, plus the weighted median calculation, (17) is positive, even if k is substantially less than n−k.

On the other hand if N and M (the non−startup iterations required by BS and SS respectively) are comparable, that phase of SS will be cheaper than the corresponding one for BS when $k \geq n/2$. Putting these observations together, one expects SS to outperform BS even when $k < n/2$, and for fixed n, the advantage should increase with k.

This assertion has some support in actual computational experience. To approximate $g(t) = \sqrt{t}$ by $\sum_{j=1}^{k} c_j t^{j-1}$ using n+1 evenly spaced points $t_i = i/n$ in [0,1] one minimizes.

$$f(\underline{c}) = \sum_{i=0}^{n} \left| \sqrt{t_i} - \sum_{j=1}^{k} c_j t_i^{j-1} \right|$$

For n = 15 and k = 5, 7, 9, and 11, both the BS and SS algorithms were applied.

Table 1

Comparison of CPU times and iteration counts for LAD approximation of $\sqrt{x}$ on [0,1] by $\sum c_j t^{j-1}$ based on 16 equally spaced points $t_i = i/15$, $i = 0,\dots,15$.

	k=5		k=7		k=9		k=11	
	CPU	ITER	CPU	ITER	CPU	ITER	CPU	ITER
BS	.036	11	.044	12	.082	16	.088	12
SS	.032	7	.042	7	.046	9	.042	6

The iteration counts for BS include the k startup steps, so they measure N+k. These results do not seem to contradict the assumption that M from (16) and N from (15) are comparable. Execution timings may be misleading because they reflect features of hardware, and peculiarities of coding. Nevertheless the data in Table 1 do accord with the conjecture that SS is more efficient than BS when $k > n/3$ and the advantage improves as k increases.

Seneta and Steiger also fit regression models of the form

$$(18) \qquad Y = c_1 X_1 + \dots + c_k X_k + U$$

following the procedure of Chapter 2. Samples of size n from (18) were generated as follows. For each $i = 1,\dots,n$, successive random numbers $x_{i2}, x_{i3}, \ldots, x_{ik}$, u_i were generated. With $x_{i1} = 1$, compute $y_i = c_i x_{i1} + \ldots + c_k x_{ik} + u_i$. The c_j were taken to be $\sqrt{i}$ so the columns of $X = (x_{ij})$ would be scaled differently.

The x_{ij} and u_i were taken to be centered Pareto random numbers of index $\alpha = 1.2$, generated using the density function

$$P(t) = \alpha/[1+(t-a)^{1+\alpha}], \; t \geq a = -\alpha/(\alpha-1).$$

They have mean zero and infinite variance.

Once $(X|\underline{y})$ was generated, LAD estimates of $\underline{c}$ using BS and SS were obtained. The process was then repeated independently until 10 samples from (18) had been considered.

Table 2 contains results for $n = 10$ and Table 3 for $n = 50$. The CPU times are net of generating the samples of (18) and are accumulated over the 10 repetitions; the iteration counts are also summed.

In both tables, the iterations differ by about 10k. Since BS iterations include the startup steps, this finding again supports the assumption that M from (16) and N from (15) are comparable; over 10 repetitions one expects this difference between the total counts. Also the CPU timings suggest that

SS < BS for k > n/3 and that the difference, BS−SS increases with k.

Table 2

Comparison of CPU Times and Iteration Counts Summed Over 10 Independent Repetitions of Fitting the Model (18) for n = 10 and
X,U Distributed as Pareto (1.2)

	k=4		k=6		k=8	
	CPU	ITER	CPU	ITER	CPU	ITER
BS	.16	65	.28	86	.32	94
SS	.18	31	.16	25	.12	10

Table 3

Comparison of CPU Times and Iteration Counts Summed Over 10 Independent Repetitions of Fitting the Model (18) for n = 50 and
X,U Distributed as Pareto (1.2)

	k=18		k=22		k=26		k=30		k=34	
	CPU	ITER	CPU	ITER	CPU	ITER	CPU	ITER	CPU	ITER
BS	7.46	324	10.04	362	13.06	395	17.20	452	20.70	473
SS	6.50	180	7.16	189	7.34	166	7.92	170	8.46	158

The device of solving (1) via (3) has potential bearing on bounded, feasible LP problems. As pointed out in Chapter 6, the problem

$$\text{(19)} \qquad \begin{aligned} &\text{maximize} && \langle \underline{d}, \underline{x} \rangle \\ &\text{subject to} && \begin{cases} C\underline{x} = \underline{a} \\ \underline{x} \geq \underline{0} \end{cases} \end{aligned}$$

where C is m by n, m < n, is equivalent to a LAD fit for n + 1 points in R^{m+1}. In Cauchy form, (3), it would use a data structure of size n−m+1 by n+2, assuming the row rank of C is m. This is smaller than (19) when $m \geq (n+1)^2/(2n+3)$, or about n/2. If C is dense, the advantage in data reduction becomes apparent, along with the potential for savings in actual computation.

7.5 Other Methods

In this section we present two methods for obtaining or approximating LAD fits. The first is iteratively reweighted least squares (IRLS). It approximates the minimizer $\hat{\underline{c}}$ of

$$(1) \qquad f(\underline{c}) = \sum_{i=1}^{n} |y_i - \langle \underline{c}, \underline{x}_i \rangle|$$

by a finite sequence $\underline{c}_0, \underline{c}_1, \ldots, \underline{c}_N$. The method is non-linear and unrelated to linear programming or the algorithms mentioned in Section 2. The second is an exact, finite algorithm for Huber M-estimation due to M. Osborne (see Clark (1981)). It gives LAD and LSQ as special cases.

The IRLS method, usually attributed to Beaton and Tukey (1974), is based on repeatedly solving a weighted least squares problem to minimize.

$$(2) \qquad g(\underline{c}) = \sum_{i=1}^{n} w_i (y_i - \langle \underline{c}, \underline{x}_i \rangle)^2, \qquad w_i \geq 0.$$

This is the usual sums of squares error function, but the points $(\underline{x}_i, y_i)$ need not be equally weighted in determining the fit. The solution is

$$(3) \qquad \underline{c} = (V^T V)^{-1} V^T \underline{u}$$

where $\underline{u}_i = w_i y_i$ and $v_{ij} = w_i x_{ij}$.

Given a continuous loss function ρ, an M-estimator is a minimizer, $\hat{\underline{c}}$, of

(4) $$h(\underline{c}) = \sum_{i=1}^{n} \rho(r_i(\underline{c})),$$

where, as usual, $r_i(\underline{c}) = y_i - \langle \underline{c}, \underline{x}_i \rangle$. The normal equations for minimizing h are

$$\sum \rho'(r_i(\underline{c}))\, x_{ij} = 0, \quad j = 1,\dots,k$$

and we define

(5) $$w_i(\underline{c}) = \rho'(r_i(\underline{c}))/r_i(\underline{c}).$$

If we regard the w_i as constants, the normal equations correspond to a weighted least squares problem. IRLS would begin with an initial value of $\underline{c}_0$, the solution to (2) with $w_i = 1$ for example. This determines a new set of weights, by (5), which gives a new weighted least squares problem (2), with solution $\underline{c}_1$, etc. This fixed point iteration could terminate either if $||\underline{c}_{i+1} - \underline{c}_i||$ is small, or if $h(\underline{c}_i) - h(\underline{c}_{i+1})$ is small.

This algorithm is described more precisely as follows.

Initialize:

[1] Accept $(\underline{x}_i, y_i)$
$j \leftarrow 0$

[2] $\underline{w} \leftarrow 1$

Iteration:

[3] $u_i \leftarrow w_i y_i;\ v_{ij} \leftarrow w_i x_{ij}$

[4] $\underline{c}_j \leftarrow (V^T V)^{-1} V^T \underline{u}$

[5] $j \leftarrow j + 1$

[6] $w_i \leftarrow \rho'(r_i(\underline{c}))/[r_i(\underline{c})]$

[7] Stop or go to 3

Suppose ρ in (4) is convex, symmetric, and differentiable on $t > 0$. A result of Dutter (1975) asserts that if $\rho'(t)/t$ is bounded and decreasing on $t > 0$, then using IRLS,

$$h(\underline{c}_{j+1}) < h(\underline{c}_j)$$

unless $\underline{c}_j$ already minimizes h. If $\hat{\underline{c}}$ is the unique minimizer of h, continuity implies that IRLS converges to $\hat{c}$. In this case one could stop the algorithm in [7] by

$$\text{"STOP if } h(c_j) - h(\underline{c}_{j+1}) < \epsilon\text{"}.$$

In LAD fitting, $\rho(t) = |t|$ and IRLS would use $w_i = 1/|r_i(\underline{c}_j)|$ at the jth step. Furthermore, IRLS approximations converge to $\hat{\underline{c}} \in M$ if (1) has a unique minimizer.

Schlossmacher (1973) investigated IRLS as a possible replacement for costly general purpose LP methods then available for obtaining LAD fits. Because $r_i(\hat{\underline{c}}) = 0$ for at least k residuals, Schlossmacher deleted a point from the data set at step j of IRLS if $|r_i(\underline{c}_j)| < \epsilon$. This prevents a prohibitively large weight w_i. He terminated the algorithm when all non-deleted residuals barely changed from the previous step:

$$|r_i(\underline{c}_j) - r_i(\underline{c}_{j+1})| < \epsilon, \text{ all } i.$$

On three small examples, the cost of IRLS approximations seemed far less than that of obtaining LAD solutions via LP. For example a 36 point fit in R^3 took 7 IRLS iterations and 1.45 CPU seconds compared to 18.41 CPU seconds to solve an LP formulation of this problem [Fisher (1961)] using an IBM library simplex routine.

However Armstrong and Frome (1976) show that Schlossmacher's optimism for IRLS was perhaps a little hasty. Their weighted median-improved BR algorithm was faster than a specially coded version of IRLS in about 98% of the random two parameter regression fits they tested. Furthermore their study showed up the possibility that in some cases, IRLS can converge very slowly indeed. Over 100 random samples from a moderate size regression model, the minimum, median, and maximum IRLS iterations were 32, 104, 1651, respectively: Hard problems caused the method great difficulty! In addition, numerical instability and convergence to wrong answers is also possible [see Fair and Peck (1974) and Gallant and Gerig (1974)]. In view of these obstacles, and with the existence of fast, exact, numerically stable algorithms like BCS, iteratively reweighted least squares cannot be considered to be a serious alternative.

However for general M-estimation, IRLS has been one of the three main competitors. Holland and Welsch (1977) focused on IRLS over two other iterative procedures for computing M-estimators. In the special case where ρ in (4) is

$$\rho_t(u) = \begin{cases} u^2/2 & \text{if } |u| \le t \\ t|u| - t^2/2 & \text{if } |u| > t \end{cases} \tag{6}$$

$t > 0$ given, the minimizer, $\underline{c}_t$, of

$$h_t(\underline{c}) = \sum_{i=1}^{n} \rho_t(r_i(\underline{c})) \tag{7}$$

is a Huber M−estimator. Let $\underline{c}_{LSQ}$ be the least squares fit to the $(\underline{x}_i, y_i)$, namely

$$\underline{c}_{LSQ} = (X^T X)^{-1} X^T \underline{y}, \tag{8}$$

and $T = \max(|r_i(\underline{c}_{LSQ})|)$. For all $t \ge T$, ρ_t treats all residuals quadratically, and $\underline{c}_t = \underline{c}_{LSQ}$. Let $\underline{c}_{LAD}$ minimize (1) and write $\tau = \min(|r_i(\underline{c}_{LAD})|: r_i \ne 0)$. For all $t \le \tau$, ρ_t treats non−zero residuals linearly and $\underline{c}_t = \underline{c}_{LAD}$. We now know how to find $\underline{c}_t$ for $t \ge T$ and $t \le \tau$.

Given t and the optimal fit, $\underline{c}_t$, write $\mathbf{S} = \{i: |r_i(\underline{c}_t)| \le t\}$, $\mathbf{N} = \{i: r_i(\underline{c}_t) < -t\}$ and $\mathbf{P} = \{i: r_i(\underline{c}_t) > t\}$. These partition the points into those with <u>s</u>mall residuals, <u>n</u>egative large residuals, and, <u>p</u>ositive large residuals, respectively. They form the <u>optimal partition</u>

If we only knew the sets **S** and **P** at the optimum, $\underline{c}_t$ could be recovered from the fact that

$$(9) \qquad h_t(\underline{c}) = \sum_S (r_i(\underline{c}))^2 + \sum_{S'} (t|r_i(\underline{c})| - t^2/2)$$

Normal equations are obtained from (9) by setting equal to zero the partial derivatives of h_t with respect to the components of $\underline{c}$. Thus

$$(10) \qquad \sum_S r_i(\underline{c})\, x_{ij} + t\,(\sum_P x_{ij} - \sum_N x_{ij}) = 0,\ j = 1,\ldots,k$$

or in matrix form,

$$(11) \qquad (X_S)^T(\underline{y}_S - X_S\underline{c}) + t[(X_P)^T - (X_N)^T]\,\underline{e} = \underline{0} \in R^k$$

where $\underline{e}$ is an n-vector of 1's and, for example, X_S has rows $\underline{x}_i$, $i \in S$. Solving (11) yields

$$(12) \qquad \underline{c}_t = [(X_S)^T X_S]^{-1}[(X_S)^T \underline{y}_S + t((X_P)^T - (X_N)^T)\underline{e}].$$

With (12), the search for $\underline{c}_t$ is reduced to the problem of finding the optimal partition of points into those with small or large positive residuals.

The following is a simplified description of a continuation algorithm due to M. Osborne. It gives the Huber M-estimators for every $t > 0$. To start, obtain

$$\underline{c}_{LSQ} = (X^T X)^{-1} X^T \underline{y}$$

and

$$t_0 = \max(|r_i(\underline{c}_{LSQ})|).$$

For $t \geq t_0$, $\underline{c}_t = \underline{c}_{LSQ}$, and $\mathbf{S} = \{1,\ldots,n\}$ is the optimal partition. To reduce t below t_0 suppose $t_0 = |r_m(\underline{c}_{LSQ})|$. Point $m \in \mathbf{S}$ is called <u>tight</u>. Delete it from **S** and include it in **P** or **N**, depending on sign $(r_m(\underline{c}_{LSQ}))$.

Now notice that (12) implies $\underline{c}_t$ is linear in t. Therefore so is each residual $r_i(\underline{c}_t) = y_i - \langle \underline{c}_t, \underline{x}_i \rangle$. As t is reduced from t_0 each residual $r_i(\underline{c}_t)$ changes at a constant rate. At some point $t_1 < t_0$ a new residual in **S** becomes tight, say $|r_p(\underline{c}_{t_1})| = t_1$. For all t in the interval $[t_1, t_0]$ the current partition is optimal and $\underline{c}_t$ is computed from (12). To continue reducing t below t_1, delete point p from **S** and include it in **P** or **N**, depending on sign $(r_p(\underline{c}_{t_1}))$.

In general, after step j, residual q has just become tight

at t_j. The current partition is optimal for all t in $[t_j, t_{j-1}]$ and for those t, $\underline{c}_t$ is computed from (12). To reduce t below t_j, if q was in **S**, delete it and include it in **P** or **N** depending on sign. On the other hand if q was in **P** or **N**, remove it and include it in S. Large residuals can become small as t increases.

At some point, $t_N > 0$, the new tight residual modifies the current partition and then $\underline{c}_t = \underline{c}_{t_N}$ for $t < t_N$. This must be the LAD fit and $((X_P)^T - (X_N)^T)\underline{e} = \underline{0}$ in (12).

To decide which residual will next become tight we need to compute $dr_i(\underline{c}_t)/dt$. From (12) the change of $\underline{c}_t$ with respect to t is given by

$$(13) \qquad \underline{c}_t' = [(X_S)^T X_S]^{-1}((X_P)^T - (X_N)^T)\underline{e},$$

so dr_i/dt is $<\underline{c}_t', \underline{x}_i>$. The largest value in $(0,t)$ where the line through $(t, r_i(\underline{c}_t))$ with slope $<\underline{c}_t', \underline{x}_i>$ hits the lines $y = x$ or $y = -x$ determines a point s_i less that t at which $r_i(\underline{c}_t)$ will first become tight as in the following diagram

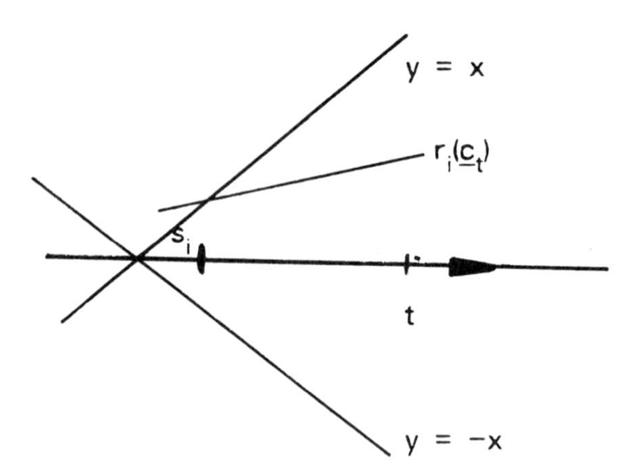

For each $i \in S$, $s_i \in (0,t)$. If $i \in \mathbf{P}$ or $i \in \mathbf{N}$ it is possible that $s_i < 0$ or that there are two values of s in (0,t) at which the ith residual becomes tight. If so, take the larger.

The continuation algorithm divides the positive reals into subintervals $(0,t_N]$, $[t_N,t_{N-1}]$,..., $[t_1,t_0]$, $[t_0,\infty)$. Each subinterval has an optimal partition which the algorithm finds. For any t in that interval, $\underline{c}_t$ may be obtained from the partition, using (12). Here is a brief description.

<u>**Input and Initialize**</u>:

[1] Accept $(\underline{x}_i, y_i) \in R^{k+1}$

[2] $j \leftarrow 0$

[3] $\underline{c}_{LSQ} \leftarrow (X^TX)^{-1}X^T\underline{y}$

[4] $\mathbf{S} \leftarrow \{1,\ldots,n\}$

[5] $t_o \leftarrow |r_m(\underline{c}_{LSQ})| = \max(|\underline{r}_i(\underline{c}_{LSQ})|)$

Reduce t Until a New Residual Is Tight:

[6] If m was in **S**, delete it and put it in **P** or **N**, depending on sign $(r_m(\underline{c}_{t_j}))$
If m was in **P** or **N**, delete it and put it in **S**.

[7] Calculate s_i, the largest point less than t_j where $r_i(\underline{c}_t)$ becomes tight, using (13).

[8] $j \leftarrow j + 1$; $t_j = \max(s_i, i \neq m) = s_p$, say.
If $t_j < 0$ STOP.

[9] set $m \leftarrow p$ and go to 6.

There may be complications. Several residuals could become tight simultaneously, for example. Clark (1981) developed a finite algorithm that can effectively handle the difficulties. The structure of optimal partitions and how h_t and minimizers depend on them is surprisingly complex. [see

Clark (1983)]. Residuals can move from **S** to **P** or **N** and back again as t decreases. There can also be different optimal partitions for the same value of t giving distinct minimizers $\underline{c}_t \neq \underline{d}_t$ of (4). This can happen if and only if the data admits non-unique LAD fits. A forthcoming paper [Clark and Osborne] will describe details of algorithms that can handle the complications which may arise in the general situation.

Finally it may be informative to show the behavior of the current algorithm on the 21 observation stack loss data set [Daniel and Wood (1980), p.61] The following table is self-explanatory.

		$\underline{c}(t_j)$				
j	t_j	c_o	c_1	c_2	c_3	$\lvert S \rvert$
0	7.23	-39.92	.716	1.295	-.152	20
1	5.82	-40.66	.750	1.201	-.143	19
9	1.26	-38.15	.838	.665	-.107	11
16	0	-39.69	.832	.574	-.061	4

The fit at step j=0 is least squares and the final fit at j=16 is LAD. The j=9 fit is one which would be recommended by Andrews (1974) based on scaling considerations for the residuals. After deleting some outliers, Daniel and Wood arrive at a least squares fit most closely resembling LAD amongst the Huber M-estimators presented in the table.

7.6 Notes

1. Karst (1958) presented an algorithm for LAD fits through the centroid of data $(x_i, y_i) \in R^2$. He suggested computation of the weighted median by sorting y_i/x_i and testing $\Sigma\ |x_i|$. For general fits he suggested the procedure mentioned in Section 1.2, degeneracy presenting a problem. Sadovski (1974) pointed this out in describing and testing an implementation of the procedure.

2. For about twenty years, from 1955 through 1975, elucidations and improvements were advanced for LAD fitting via LP. Davies (1967) eg., described a primal formulation, incorrectly asserted that the two problems were equivalent, and worked an example. No references to previous work were given. Robers and Ben-Israel (1969) enunciated a general algorithm, SUBOPT, for the bounded variable, or interval linear programming problem and applied it to the bounded variable dual form (6.2.5) of a LAD fit. It could be regarded as constituting a general framework within which the algorithms of Section 2 fall.

The work of Wagner (1959), Rabinowitz (1965) and Robers and Ben-Israel (1969) suggested that the bounded variable, dual form was the appropriate LP problem when n was much larger than k. However Barrodale and Young (1966) showed how to store a primal LP form efficiently and Barrodale and Roberts (1970) suggested an efficient startup leading to an extreme fit. These contributions paved the way for the multiple pivot sequences of Barrodale and Roberts (1973) and the BR algorithm. This primal LP variant was the most efficient procedure of its time, contrary to the belief in the dual form. Spyropoulos, Kiountouzis, and Young (1973) also describe a primal LP algorithm with multiple pivot sequences and report large savings over Robers and Ben-Israel, which was then regarded as the leading method prior to BR. It is ALGOL coded and seems not to have been compared with the Section 2 procedures.

Abdelmalek (1974) proposed solving (6.2.5) by the dual simplex algorithm. Later [Abdelmalek (1975)] he introduced multiple pivot sequences and asserted that this dual formulation was comparable to BR over a variety of test problems. This is not surprising since the two algorithms are identical [Armstrong and Godfrey (1979)] except for the initial solution and the way in which ties are resolved.

Earlier, Usow (1967) had described a descent method for LAD approximation of a function g by $c_1\phi_1 + \dots + c_k\phi_k$ over points $t_1,\dots,t_n$, the ϕ_i forming a Tchebycheff set. Abdelmalek (1974) "generalized" Usow's results to LAD fitting of n points in R^{k+1} and showed that Usow's algorithm was in fact the dual simplex applied to (6.2.5), several iterations of the latter corresponding to one of the former. He later showed that Abdelmalek (1975) was completely identical to Usow.

Narula and Wellington (1973), (1977a), (1977b), (1977c) have a series of papers on LAD fitting through the centroid of $(\underline{x}_i, y_i)$ using the dual simplex algorithm. They seem unaware of BR, and the work of Abdelmalek, Robers and Ben-Israel, Usow and others on the dual formulation. The algorithm (1977a) may be incorrect anyway. In (1977b) they describe an algorithm to minimize $\sum |y_i - \langle \underline{c}, \underline{x}_i \rangle| / |y_i|$ and compare the results of using it with their LAD procedure from (1977a). Bloomfield and Steiger (1980b) point out some errors in this work.

3. Bartels, Conn and Sinclair (1978) asserted that their algorithm, though obviously similar to BR,was not identical. Their demonstration used a simple problem, starting both algorithms from the same extreme fit. The OBSERVATION in Subsection 7.2.2 didn't apply to this example because BR was still in its startup phase.

4. Bartels, Conn, and Sinclair (1977, 1978) suggested the projected gradient idea for choosing a descent directions in the minimization of piecewise differentiable functions.

5. McCormick and Sposito (1976) suggests that if the least squares fit to $(\underline{x}_i, y_i)$ is already known, it may be expedient to use it in starting up LAD algorithms.

6. Constrained LAD fitting algorithms were studied in Armstrong and Hultz (1977) and Barrodale and Roberts (1977, 1978).

7. Barrodale and Roberts (1972) tested BR against the SUBOPT and Usow methods and found it superior. Armstrong, Frome, and Kung (1979) modified BR with "a one-pass partial sort routine" and an implementation using an LU decomposition. Their test results, while difficult to interpret, suggest that BR with an efficient weighted median computation is at least comparable to the other methods. Gilsinn, et. al. (1977) compare BR, BCS, and Armstrong and Frome (1976). The results are difficult to interpret because their Figures 2 and 3 disagree with their Table III as well as with the results of our Section 4. Their Table 3 is not inconsistent with the findings in our Section 4.

8. APPENDIX

In this section we present the detailed results of the Monte-Carlo study described in Section 2.4. Samples of size n were generated from

$$Y = 1 + X_1 + X_2 + U \tag{1}$$

where the n values of X_1 were

$$a_i = [i - (n+1)/2]/[(n^2-1)/12]^{1/2}, \quad i=1,\ldots n.$$

The values of X_2 were n independently generated standard exponentials, centered at 0. The process was repeated independently, 901 times. For each repetition, the coefficients $(1,1,1)'$ were estimated by LAD, LSQ, and the two Huber M-estimators, here denoted by H-05 and H-25. The error distributions of the four estimators over the 901 repetitions are summarized in the following tables.

There is a table for each distribution used for U in (1), eight in all, and for each sample size n=10,50,100; thus, there are 24 tables. The distributions used for U are described in Section 2.4. They are normal, the two Huber least-informative distributions, here denoted H.05 and H.25, double exponential, the two pareto contaminated distributions, here denoted P.05 and P.25, the pure pareto(1.2), here denoted PARETO, and the logistic. For a detailed description of these densities, refer to Section 2.4.

In each table the components of the error of estimation

$$\underline{e} = \underline{c} - (1,1,1)'$$

are described for each estimator, $\underline{c}$, as well as the length $|\underline{e}|$. The following percentiles of the 901 estimation errors are given: minimum, fifth (x 05), 25th (Q1), median (Q2), 75th (Q3), 95th (x 95), and the maximim. In addition the mean, standard deviation (S.D.), and interquartile range (IQR) are given.

TABLE

Estimation errors in Regressions (2.5.8), NORMAL Errors, 901 Repetitions with Sample Size = 10

LSQ

component of e

	1	2	3	|e|
min.	-1.4316	-1.4362	-2.7009	0.0841
x 05	-0.5381	-0.5316	-0.7912	0.2107
Q1	-0.2207	-0.2307	-0.2747	0.3982
Q2	0.0190	-0.0044	-0.0031	0.5510
Q3	0.2536	0.2224	0.2706	0.7713
x 95	0.6020	0.5301	0.7738	1.2060
max.	1.0282	1.3165	1.9696	2.9420
MEAN	0.0189	-0.0042	-0.0009	0.6152
S.D.	0.3504	0.3413	0.4982	0.3301
IQR.	0.4742	0.4531	0.5453	0.3731

H-05

component of e

	1	2	3	|e|
min.	-2.0192	-1.1842	-2.5996	0.0933
x 05	-0.5533	-0.5412	-0.8190	0.2075
Q1	-0.2234	-0.2334	-0.2672	0.3926
Q2	0.0062	-0.0140	-0.0063	0.5604
Q3	0.2554	0.2313	0.2706	0.7863
x 95	0.5947	0.5572	0.8127	1.2257
max.	1.0282	1.3388	1.9696	3.4038
MEAN	0.0155	-0.0036	-0.0016	0.6216
S.D.	0.3557	0.3454	0.5037	0.3361
IQR.	0.4788	0.4647	0.5378	0.3936

H-25

component of e

	1	2	3	|e|
min.	-2.4183	-1.1620	-2.7458	0.0768
x 05	-0.5895	-0.5785	-0.8195	0.2121
Q1	-0.2391	-0.2542	-0.2640	0.4094
Q2	0.0104	-0.0099	0.0055	0.5789
Q3	0.2708	0.2468	0.2661	0.8149
x 95	0.6190	0.5686	0.8712	1.2754
max.	0.9771	1.3283	2.1104	3.7617
MEAN	0.0123	-0.0044	0.0031	0.6438
S.D.	0.3710	0.3606	0.5163	0.3455
IQR.	0.5098	0.5010	0.5301	0.4054

LAD

component of e

	1	2	3	|e|
min.	-2.8865	-1.3337	-2.8173	0.0480
x 05	-0.6717	-0.6643	-0.9398	0.2268
Q1	-0.2583	-0.2840	-0.2861	0.4671
Q2	0.0238	-0.0095	0.0307	0.6693
Q3	0.3184	0.2665	0.3333	0.9393
x 95	0.7383	0.6824	0.9842	1.4249
max.	1.2297	1.6571	2.1105	4.1132
MEAN	0.0304	-0.0014	0.0205	0.7348
S.D.	0.4317	0.4112	0.5828	0.3950
IQR.	0.5767	0.5505	0.6195	0.4722

TABLE

Estimation errors in Regressions (2.5.8), H.05
Errors, 901 Repetitions with Sample Size = 10

LSQ

component of e

	1	2	3	|e|
min.	-2.9424	-1.6511	-4.4949	0.0512
x 05	-0.5546	-0.6295	-0.8235	0.1963
Q1	-0.2406	-0.2237	-0.2538	0.3898
Q2	-0.0193	0.0217	0.0447	0.5671
Q3	0.2200	0.2529	0.2931	0.7780
x 95	0.5959	0.6152	0.7906	1.3100
max.	1.1338	1.1406	2.6523	5.3776
MEAN	-0.0130	0.0075	0.0150	0.6372
S.D.	0.3700	0.3782	0.5379	0.4041
IQR.	0.4606	0.4766	0.5469	0.3882

H-05

component of e

	1	2	3	|e|
min.	-2.9424	-1.2575	-4.4949	0.0512
x 05	-0.5341	-0.6055	-0.8176	0.1910
Q1	-0.2399	-0.2247	-0.2538	0.3936
Q2	-0.0211	0.0183	0.0363	0.5622
Q3	0.2200	0.2489	0.2990	0.7753
x 95	0.5726	0.5796	0.7452	1.2369
max.	1.1338	1.1406	2.7191	5.3776
MEAN	-0.0126	0.0063	0.0100	0.6261
S.D.	0.3653	0.3661	0.5265	0.3907
IQR.	0.4600	0.4735	0.5528	0.3816

H-25

component of e

	1	2	3	|e|
min.	-2.5206	-1.1469	-3.9852	0.0712
x 05	-0.5630	-0.6177	-0.8143	0.2094
Q1	-0.2650	-0.2530	-0.2759	0.4248
Q2	-0.0115	0.0144	0.0318	0.5941
Q3	0.2335	0.2445	0.3074	0.7959
x 95	0.6221	0.6311	0.8064	1.2721
max.	1.3221	1.1330	2.4289	4.6019
MEAN	-0.0102	0.0029	0.0102	0.6604
S.D.	0.3859	0.3723	0.5481	0.3893
IQR.	0.4985	0.4974	0.5833	0.3711

LAD

component of e

	1	2	3	|e|
min.	-3.4844	-1.0835	-4.6701	0.0530
x 05	-0.7139	-0.6876	-0.9781	0.2528
Q1	-0.3116	-0.2979	-0.3127	0.5107
Q2	-0.0153	-0.0060	0.0264	0.6886
Q3	0.2941	0.2781	0.3394	0.9293
x 95	0.7268	0.7158	0.9507	1.4716
max.	1.6456	1.3599	3.0219	5.8274
MEAN	-0.0149	-0.0017	0.0046	0.7659
S.D.	0.4637	0.4182	0.6385	0.4589
IQR.	0.6057	0.5760	0.6520	0.4186

TABLE

Estimation errors in Regressions (2.5.8), H.25
Errors, 901 Repetitions with Sample Size = 10

LSQ

	component of e			
	1	2	3	|e|
min.	-2.3757	-1.6413	-4.6553	0.0387
x 05	-0.8894	-0.7840	-1.0598	0.1940
Q1	-0.2823	-0.2693	-0.3042	0.4232
Q2	-0.0047	0.0105	0.0202	0.6543
Q3	0.2915	0.2643	0.3040	1.0082
x 95	0.7677	0.7401	1.0503	1.7838
max.	2.0545	2.5923	3.9447	5.2969
MEAN	-0.0037	-0.0019	0.0069	0.7872
S.D.	0.4926	0.4681	0.6831	0.5551
IQR.	0.5738	0.5336	0.6083	0.5850

H-05

	component of e			
	1	2	3	|e|
min.	-2.9833	-1.2043	-4.6723	0.0387
x 05	-0.7121	-0.6100	-0.9626	0.1705
Q1	-0.2213	-0.2353	-0.2641	0.3344
Q2	0.0050	-0.0021	0.0018	0.5414
Q3	0.2230	0.2104	0.2588	0.8329
x 95	0.6341	0.6040	0.8959	1.5789
max.	1.8396	1.9776	3.3592	5.3435
MEAN	-0.0088	-0.0024	0.0061	0.6735
S.D.	0.4203	0.3869	0.6306	0.5197
IQR.	0.4443	0.4457	0.5229	0.4985

H-25

component of e

	1	2	3	|e|
min.	-3.2497	-1.5036	-4.7794	0.0291
x 05	-0.6495	-0.5896	-0.9016	0.1596
Q1	-0.1903	-0.2106	-0.2338	0.2985
Q2	0.0013	0.0126	-0.0134	0.4831
Q3	0.1915	0.1927	0.2361	0.7851
x 95	0.6214	0.5588	0.9031	1.5747
max.	1.8035	2.1239	3.4749	5.5411
MEAN	-0.0067	0.0008	0.0023	0.6266
S.D.	0.4034	0.3638	0.6153	0.5298
IQR.	0.3818	0.4033	0.4699	0.4866

LAD

component of e

	1	2	3	|e|
min.	-2.7022	-1.8094	-5.2854	0.0379
x 05	-0.6544	-0.6020	-0.8527	0.1477
Q1	-0.2096	-0.1921	-0.2358	0.3152
Q2	0.0093	0.0091	-0.0046	0.4744
Q3	0.2038	0.2183	0.2513	0.8176
x 95	0.6525	0.6565	0.9918	1.6289
max.	2.2273	2.1617	3.5086	6.1071
MEAN	0.0016	0.0129	0.0067	0.6443
S.D.	0.4189	0.3844	0.6180	0.5384
IQR.	0.4134	0.4104	0.4871	0.5024

TABLE

Estimation errors in Regressions (2.5.8), DEXP Errors, 901 Repetitions with Sample Size = 10

LSQ

	component of e			
	1	2	3	|e|
min.	-3.1546	-1.4913	-5.3396	0.0572
x 05	-0.8734	-0.7632	-1.1036	0.2463
Q1	-0.3211	-0.2992	-0.3497	0.4993
Q2	-0.0059	-0.0134	-0.0096	0.7289
Q3	0.3189	0.2982	0.3294	1.0471
x 95	0.8306	0.8420	1.0575	1.7992
max.	1.4987	2.3204	4.3705	6.2373
MEAN	-0.0137	0.0104	-0.0147	0.8447
S.D.	0.5256	0.4958	0.7037	0.5511
IQR.	0.6400	0.5975	0.6791	0.5479

H-05

	component of e			
	1	2	3	|e|
min.	-3.5568	-1.6142	-5.7981	0.0572
x 05	-0.8179	-0.7051	-1.1068	0.2328
Q1	-0.3051	-0.2656	-0.3114	0.4676
Q2	-0.0037	-0.0006	0.0009	0.6885
Q3	0.2975	0.2923	0.3178	0.9737
x 95	0.7666	0.7716	0.9962	1.7109
max.	1.7365	1.9518	4.3080	6.8241
MEAN	-0.0137	0.0170	-0.0119	0.7966
S.D.	0.4992	0.4646	0.6698	0.5282
IQR.	0.6026	0.5578	0.6292	0.5061

H-25

component of e

	1	2	3	|e|
min.	-3.7684	-1.6804	-5.8703	0.0580
x 05	-0.8026	-0.7252	-1.0521	0.2245
Q1	-0.2892	-0.2672	-0.3190	0.4460
Q2	-0.0062	0.0043	0.0035	0.6765
Q3	0.2996	0.3036	0.3315	0.9629
x 95	0.7504	0.8111	1.0031	1.7502
max.	2.1640	2.1124	4.2685	6.9936
MEAN	-0.0114	0.0204	-0.0082	0.7909
S.D.	0.4971	0.4687	0.6782	0.5488
IQR.	0.5889	0.5708	0.6504	0.5169

LAD

component of e

	1	2	3	|e|
min.	-3.8539	-1.7466	-6.0960	0.0612
x 05	-0.8752	-0.8098	-1.1939	0.2524
Q1	-0.3020	-0.2932	-0.3326	0.4529
Q2	-0.0129	0.0210	0.0015	0.7178
Q3	0.2988	0.3398	0.3456	1.0545
x 95	0.7918	0.8831	1.0983	1.9428
max.	2.5752	2.0809	3.4286	7.2564
MEAN	-0.0183	0.0291	-0.0098	0.8525
S.D.	0.5351	0.5087	0.7397	0.6049
IQR.	0.6009	0.6330	0.6783	0.6016

TABLE

Estimation errors in Regressions (2.5.8), P.05
Errors, 901 Repetitions with Sample Size = 10

LSQ

component of e

	1	2	3	|e|
min.	-115.6722	-71.8131	-32.4531	0.0473
x 05	-1.1995	-1.3454	-1.4639	0.2660
Q1	-0.3824	-0.3774	-0.4617	0.5557
Q2	0.0092	-0.0048	-0.0511	0.8871
Q3	0.4059	0.3419	0.4377	1.4931
x 95	1.4648	1.1362	1.6508	4.0494
max.	37.1821	172.7842	28.4308	208.1945
MEAN	-0.1493	0.1123	-0.0527	1.7809
S.D.	4.6695	6.4253	2.3762	8.0992
IQR.	0.7882	0.7193	0.8993	0.9374

H-05

component of e

	1	2	3	|e|
min.	-2.5385	-1.8910	-4.2605	0.0473
x 05	-0.8973	-0.8603	-1.2360	0.2265
Q1	-0.3174	-0.2907	-0.3759	0.4668
Q2	-0.0017	0.0145	-0.0269	0.7295
Q3	0.3122	0.3056	0.3704	1.1109
x 95	0.9086	0.8015	1.3068	2.1798
max.	4.6625	4.6710	7.0457	8.4720
MEAN	0.0124	0.0140	-0.0005	0.9091
S.D.	0.5865	0.5423	0.8488	0.7291
IQR.	0.6296	0.5963	0.7463	0.6440

H-25

component of e

	1	2	3	|e|
min.	-2.5233	-1.8982	-4.4044	0.0434
x 05	-0.8334	-0.8009	-1.2480	0.2286
Q1	-0.2954	-0.2855	-0.3461	0.4726
Q2	-0.0060	0.0054	-0.0245	0.6944
Q3	0.3006	0.2896	0.3619	1.0783
x 95	0.8582	0.7781	1.3063	2.0926
max.	5.0732	1.8562	7.2135	8.8326
MEAN	0.0132	0.0099	0.0111	0.8792
S.D.	0.5685	0.4988	0.8351	0.7042
IQR.	0.5960	0.5751	0.7079	0.6057

LAD

component of e

	1	2	3	|e|
min.	-2.6487	-1.8257	-4.7558	0.0756
x 05	-0.8721	-0.8189	-1.2752	0.2460
Q1	-0.3257	-0.3076	-0.3835	0.4973
Q2	-0.0104	0.0234	-0.0100	0.7477
Q3	0.3175	0.3139	0.3998	1.1412
x 95	0.8708	0.8778	1.3676	2.2486
max.	5.5554	2.1602	7.5087	9.3430
MEAN	0.0130	0.0144	0.0121	0.9426
S.D.	0.6097	0.5278	0.8910	0.7450
IQR.	0.6433	0.6215	0.7832	0.6439

TABLE

Estimation errors in Regressions (2.5.8), P.25
Errors, 901 Repetitions with Sample Size = 10

LSQ

	component of e			
	1	2	3	|e|
min.	-110.9318	-79.1756	-171.7676	0.1877
x 05	-3.9041	-3.5066	-4.2792	0.6395
Q1	-0.8793	-0.8797	-1.1520	1.3514
Q2	0.0506	0.0831	0.0592	2.2512
Q3	1.0482	0.9604	1.1748	3.9518
x 95	5.0750	4.2399	5.4951	14.9955
max.	170.9984	183.2174	169.3517	292.0177
MEAN	0.4209	0.3470	0.4561	5.0309
S.D.	8.5985	8.1181	11.1845	15.4951
IQR.	1.9275	1.8401	2.3268	2.6004

H-05

	component of e			
	1	2	3	|e|
min.	-34.0620	-79.1756	-36.7247	0.1085
x 05	-2.1772	-2.3826	-2.8055	0.5348
Q1	-0.7493	-0.7529	-0.9497	1.1249
Q2	-0.0109	0.0030	0.0035	1.7811
Q3	0.7408	0.7060	0.9133	2.7073
x 95	2.3687	2.1889	3.2942	6.2198
max.	92.7536	35.7595	55.0812	102.2885
MEAN	0.2204	-0.0538	0.1746	2.7901
S.D.	4.0635	3.8774	3.9503	6.2801
IQR.	1.4901	1.4589	1.8630	1.5824

H-25

	component of e			
	1	2	3	|e|
min.	-17.5255	-17.7317	-36.7247	0.1057
x 05	-2.0405	-2.0238	-2.9784	0.3682
Q1	-0.7207	-0.7384	-0.9387	1.1120
Q2	0.0085	-0.0197	0.0196	1.8161
Q3	0.7285	0.7244	0.8947	2.6571
x 95	2.3059	2.1607	3.2655	5.7706
max.	27.9723	11.8604	40.6096	50.7174
MEAN	0.0751	0.0052	0.0688	2.3462
S.D.	1.8605	1.5594	2.8679	2.9356
IQR.	1.4492	1.4628	1.8334	1.5451

LAD

	component of e			
	1	2	3	|e|
min.	-17.5255	-17.7317	-36.7247	0.0874
x 05	-2.1662	-2.1219	-3.0634	0.3352
Q1	-0.7197	-0.8123	-0.9253	1.1225
Q2	0.0112	-0.0244	0.0171	1.8885
Q3	0.7486	0.7810	0.9138	2.8084
x 95	2.3605	2.2453	3.3138	5.8256
max.	27.9723	11.8604	40.6096	50.7174
MEAN	0.0688	0.0089	0.0711	2.3981
S.D.	1.8829	1.6019	2.8968	2.9590
IQR.	1.4682	1.5933	1.8391	1.6859

TABLE

Estimation errors in Regressions (2.5.8), PARETO Errors, 901 Repetitions with Sample Size = 10

LSQ

	component of e			
	1	2	3	|e|
min.	-53.5966	-60.2846	-683.0263	0.1604
x 05	-4.2750	-4.0179	-5.3811	0.4112
Q1	-0.8829	-0.7537	-1.1385	0.9623
Q2	0.0217	0.0001	-0.0630	1.9164
Q3	0.8044	0.7678	0.8144	4.1395
x 95	4.8449	4.3736	5.7609	15.9854
max.	63.0413	663.9647	44.0892	952.7501
MEAN	0.0919	0.7542	-0.8494	5.3264
S.D.	5.4826	22.6777	23.3035	32.5621
IQR.	1.6873	1.5215	1.9529	3.1771

H-05

	component of e			
	1	2	3	|e|
min.	-53.5966	-13.7387	-683.0263	0.1170
x 05	-1.6859	-1.4010	-2.8686	0.2640
Q1	-0.4126	-0.4060	-0.5643	0.5923
Q2	0.0154	0.0277	-0.0370	0.9941
Q3	0.4568	0.4104	0.4951	1.8050
x 95	1.6919	1.6180	2.2853	5.4154
max.	26.7510	663.9647	8.5473	952.7501
MEAN	-0.0062	0.8233	-0.9876	2.8989
S.D.	2.5720	22.2479	22.9503	31.9616
IQR.	0.8693	0.8164	1.0594	1.2127

H-25

component of e

	1	2	3	|e|
min.	-13.7547	-6.0421	-35.8712	0.0640
x 05	-1.4549	-1.3171	-2.4186	0.2443
Q1	-0.3679	-0.3582	-0.4985	0.5151
Q2	-0.0010	0.0156	0.0144	0.9010
Q3	0.4043	0.3575	0.4586	1.6773
x 95	1.5201	1.3995	2.1738	4.5378
max.	6.3888	28.6808	11.8283	46.0434
MEAN	0.0079	0.0380	-0.1129	1.5128
S.D.	1.1549	1.3550	2.1812	2.3770
IQR.	0.7722	0.7158	0.9572	1.1622

LAD

component of e

	1	2	3	|e|
min.	-13.7547	-6.2954	-35.8712	0.0351
x 05	-1.5050	-1.2153	-2.4568	0.2067
Q1	-0.3889	-0.3483	-0.5202	0.4874
Q2	0.0040	0.0232	0.0140	0.9031
Q3	0.3703	0.3566	0.4496	1.6939
x 95	1.5577	1.5218	2.1164	4.5349
max.	6.3888	28.6808	11.8283	46.0434
MEAN	0.0033	0.0678	-0.1083	1.5140
S.D.	1.1759	1.3494	2.1953	2.3968
IQR.	0.7592	0.7049	0.9697	1.2065

TABLE

Estimation errors in Regressions (2.5.8), LOGISTIC Errors, 901 Repetitions with Sample Size = 10

LSQ

	component of e			
	1	2	3	|e|
min.	-2.7910	-1.9493	-4.7063	0.0450
x 05	-1.0201	-1.0205	-1.5378	0.3520
Q1	-0.4161	-0.3917	-0.5201	0.6793
Q2	0.0086	0.0075	-0.0634	0.9816
Q3	0.4099	0.4071	0.4137	1.3543
x 95	1.0183	1.0504	1.3234	2.1906
max.	2.0917	2.3444	4.4202	5.2511
MEAN	0.0028	-0.0028	-0.0655	1.0919
S.D.	0.6341	0.6219	0.8804	0.6120
IQR.	0.8259	0.7988	0.9338	0.6750

H-05

	component of e			
	1	2	3	|e|
min.	-2.1192	-2.8863	-8.0472	0.0929
x 05	-1.0343	-1.1146	-1.4606	0.3678
Q1	-0.3952	-0.3897	-0.5400	0.6781
Q2	0.0257	0.0114	-0.0570	0.9957
Q3	0.4083	0.3912	0.4526	1.3831
x 95	1.0652	1.0286	1.2832	2.1909
max.	1.9747	2.3162	4.6698	8.7528
MEAN	0.0075	-0.0121	-0.0568	1.1059
S.D.	0.6346	0.6486	0.9104	0.6568
IQR.	0.8035	0.7808	0.9926	0.7051

H-25

component of e

	1	2	3	|e|
min.	-3.2136	-3.3408	-9.8220	0.1067
x 05	-1.0766	-1.1310	-1.6685	0.3738
Q1	-0.4226	-0.4156	-0.5689	0.7045
Q2	0.0197	0.0061	-0.0521	1.0436
Q3	0.4407	0.4178	0.4860	1.4514
x 95	1.1253	1.0721	1.3458	2.3406
max.	1.9245	2.3947	5.2994	10.6346
MEAN	0.0076	-0.0067	-0.0692	1.1653
S.D.	0.6714	0.6776	0.9774	0.7147
IQR.	0.8633	0.8334	1.0548	0.7470

LAD

component of e

	1	2	3	|e|
min.	-3.2200	-2.5312	-6.3989	0.1136
x 05	-1.2250	-1.2571	-1.8528	0.3793
Q1	-0.4600	-0.4829	-0.6086	0.7765
Q2	0.0037	0.0134	-0.0562	1.1407
Q3	0.4763	0.4423	0.5074	1.5883
x 95	1.2007	1.2287	1.4889	2.5605
max.	2.5170	2.7263	6.0813	6.9104
MEAN	0.0033	0.0033	-0.0747	1.2764
S.D.	0.7476	0.7341	1.0384	0.7421
IQR.	0.9363	0.9252	1.1161	0.8118

TABLE

Estimation errors in Regressions (2.5.8), NORMAL Errors, 901 Repetitions with Sample Size = 50

LSQ

component of e

	1	2	3	|e|
min.	-0.5007	-0.5276	-0.4087	0.0159
x 05	-0.2543	-0.2276	-0.2344	0.0784
Q1	-0.1117	-0.0867	-0.0995	0.1552
Q2	-0.0044	0.0037	0.0001	0.2265
Q3	0.0898	0.0974	0.0939	0.2944
x 95	0.2268	0.2484	0.2372	0.4035
max.	0.6217	0.5343	0.6693	0.9276
MEAN	-0.0094	0.0062	-0.0018	0.2313
S.D.	0.1479	0.1431	0.1473	0.1031
IQR.	0.2014	0.1841	0.1933	0.1393

H-05

component of e

	1	2	3	|e|
min.	-0.4687	-0.5167	-0.4711	0.0057
x 05	-0.2613	-0.2321	-0.2566	0.0794
Q1	-0.1214	-0.0878	-0.1024	0.1584
Q2	-0.0052	0.0015	-0.0021	0.2317
Q3	0.0897	0.1015	0.0966	0.3049
x 95	0.2321	0.2525	0.2522	0.4192
max.	0.6176	0.5709	0.5660	0.8502
MEAN	-0.0113	0.0072	-0.0013	0.2377
S.D.	0.1513	0.1471	0.1515	0.1054
IQR.	0.2110	0.1893	0.1990	0.1465

H-25

component of e

	1	2	3	|e|
min.	-0.4917	-0.4934	-0.4753	0.0108
x 05	-0.2845	-0.2339	-0.2623	0.0872
Q1	-0.1245	-0.0972	-0.1032	0.1714
Q2	-0.0035	0.0066	-0.0021	0.2410
Q3	0.0962	0.1014	0.1070	0.3400
x 95	0.2439	0.2547	0.2632	0.4461
max.	0.6006	0.5723	0.5481	0.7795
MEAN	-0.0133	0.0071	0.0003	0.2498
S.D.	0.1589	0.1540	0.1595	0.1103
IQR.	0.2207	0.1986	0.2102	0.1486

LAD

component of e

	1	2	3	|e|
min.	-0.5787	-0.5319	-0.6119	0.0184
x 05	-0.3145	-0.2819	-0.3084	0.1081
Q1	-0.1324	-0.1221	-0.1135	0.2015
Q2	-0.0138	0.0041	-0.0034	0.2735
Q3	0.1064	0.1328	0.1150	0.3676
x 95	0.2750	0.2954	0.3086	0.5131
max.	0.5857	0.5913	0.5600	0.7636
MEAN	-0.0142	0.0053	0.0000	0.2898
S.D.	0.1784	0.1826	0.1841	0.1234
IQR.	0.2388	0.2549	0.2285	0.1661

TABLE

Estimation errors in Regressions (2.5.8), H.05
Errors, 901 Repetitions with Sample Size = 50

LSQ

	component of e			
	1	2	3	|e|
min.	-0.5471	-0.5242	-0.5606	0.0094
x 05	-0.2433	-0.2417	-0.2366	0.0788
Q1	-0.0892	-0.0961	-0.0920	0.1597
Q2	0.0096	0.0026	0.0150	0.2252
Q3	0.1123	0.1061	0.1043	0.3036
x 95	0.2475	0.2397	0.2842	0.4458
max.	0.4456	0.5398	0.6333	0.7997
MEAN	0.0078	0.0024	0.0097	0.2399
S.D.	0.1482	0.1500	0.1607	0.1134
IQR.	0.2015	0.2022	0.1963	0.1439

H-05

	component of e			
	1	2	3	|e|
min.	-0.4890	-0.4645	-0.4724	0.0206
x 05	-0.2288	-0.2411	-0.2358	0.0730
Q1	-0.0887	-0.0906	-0.0895	0.1533
Q2	0.0097	0.0057	0.0087	0.2159
Q3	0.1014	0.1016	0.1050	0.2889
x 95	0.2380	0.2263	0.2576	0.4156
max.	0.4228	0.4365	0.6147	0.7352
MEAN	0.0075	0.0037	0.0091	0.2278
S.D.	0.1405	0.1429	0.1495	0.1035
IQR.	0.1902	0.1921	0.1944	0.1357

H-25

component of e

	1	2	3	|e|
min.	-0.5145	-0.4977	-0.5340	0.0206
x 05	-0.2704	-0.2641	-0.2645	0.0910
Q1	-0.1011	-0.0960	-0.1004	0.1771
Q2	0.0168	0.0065	0.0102	0.2490
Q3	0.1169	0.1165	0.1123	0.3290
x 95	0.2709	0.2660	0.2727	0.4577
max.	0.4594	0.4735	0.6628	0.7306
MEAN	0.0081	0.0061	0.0093	0.2577
S.D.	0.1610	0.1614	0.1633	0.1109
IQR.	0.2180	0.2125	0.2128	0.1519

LAD

component of e

	1	2	3	|e|
min.	-0.5826	-0.5703	-0.6491	0.0356
x 05	-0.3220	-0.3104	-0.3299	0.1060
Q1	-0.1185	-0.1188	-0.1036	0.1973
Q2	0.0058	0.0098	0.0233	0.2822
Q3	0.1373	0.1311	0.1357	0.3745
x 95	0.3038	0.3140	0.2991	0.5270
max.	0.6273	0.4934	0.8706	0.8831
MEAN	0.0053	0.0077	0.0116	0.2960
S.D.	0.1858	0.1878	0.1864	0.1305
IQR.	0.2558	0.2499	0.2393	0.1772

TABLE

Estimation errors in Regressions (2.5.8), H.25
Errors, 901 Repetitions with Sample Size = 50

LSQ

component of e

	1	2	3	|e|
min.	-0.7137	-0.5778	-0.7580	0.0251
x 05	-0.3442	-0.3235	-0.3575	0.1051
Q1	-0.1256	-0.1261	-0.1331	0.2013
Q2	0.0140	0.0022	-0.0043	0.2879
Q3	0.1233	0.1215	0.1238	0.3977
x 95	0.3087	0.3139	0.3292	0.5772
max.	0.6648	0.6112	0.7424	0.8379
MEAN	-0.0008	0.0005	-0.0067	0.3100
S.D.	0.1952	0.1923	0.2068	0.1473
IQR.	0.2489	0.2476	0.2569	0.1964

H-05

component of e

	1	2	3	|e|
min.	-0.4460	-0.4656	-0.5394	0.0185
x 05	-0.2324	-0.2226	-0.2660	0.0768
Q1	-0.0875	-0.0901	-0.0985	0.1431
Q2	0.0001	0.0005	-0.0066	0.2142
Q3	0.0843	0.0942	0.0811	0.2906
x 95	0.2272	0.2377	0.2574	0.4319
max.	0.4416	0.4417	0.7189	0.7397
MEAN	-0.0019	0.0008	-0.0056	0.2264
S.D.	0.1379	0.1413	0.1566	0.1104
IQR.	0.1718	0.1843	0.1797	0.1476

H-25

component of e

	1	2	3	|e|
min.	-0.3744	-0.4150	-0.5180	0.0076
x 05	-0.1922	-0.1893	-0.2247	0.0608
Q1	-0.0785	-0.0746	-0.0811	0.1207
Q2	-0.0023	0.0001	-0.0001	0.1760
Q3	0.0729	0.0808	0.0676	0.2457
x 95	0.1991	0.1987	0.2205	0.3679
max.	0.3874	0.3724	0.7251	0.7506
MEAN	-0.0017	0.0021	-0.0045	0.1905
S.D.	0.1150	0.1193	0.1348	0.0968
IQR.	0.1514	0.1554	0.1488	0.1251

LAD

component of e

	1	2	3	|e|
min.	-0.3532	-0.4200	-0.6316	0.0101
x 05	-0.2059	-0.1948	-0.2327	0.0547
Q1	-0.0733	-0.0753	-0.0834	0.1118
Q2	0.0013	0.0009	-0.0036	0.1725
Q3	0.0706	0.0716	0.0696	0.2529
x 95	0.1960	0.2005	0.2273	0.3835
max.	0.3569	0.4362	0.5285	0.6641
MEAN	-0.0011	0.0002	-0.0063	0.1903
S.D.	0.1175	0.1197	0.1370	0.1034
IQR.	0.1439	0.1469	0.1530	0.1412

TABLE

Estimation errors in Regressions (2.5.8), DEXP Errors, 901 Repetitions with Sample Size = 50

LSQ

	component of e			
	1	2	3	|e|
min.	-0.6566	-0.7424	-0.8314	0.0237
x 05	-0.3327	-0.3287	-0.3871	0.1134
Q1	-0.1360	-0.1285	-0.1456	0.2206
Q2	-0.0045	0.0087	-0.0117	0.3164
Q3	0.1410	0.1363	0.1343	0.4228
x 95	0.3489	0.3378	0.3810	0.6032
max.	0.6175	0.6894	0.9107	0.9699
MEAN	0.0005	0.0039	-0.0052	0.3312
S.D.	0.2038	0.1963	0.2288	0.1504
IQR.	0.2770	0.2648	0.2799	0.2023

H-05

	component of e			
	1	2	3	|e|
min.	-0.5882	-0.5455	-0.7627	0.0322
x 05	-0.2942	-0.2864	-0.3412	0.0958
Q1	-0.1217	-0.1129	-0.1264	0.1924
Q2	-0.0045	0.0050	-0.0041	0.2763
Q3	0.1185	0.1176	0.1192	0.3713
x 95	0.2990	0.2798	0.3321	0.5268
max.	0.4844	0.6574	0.7848	0.8498
MEAN	-0.0017	0.0035	-0.0037	0.2904
S.D.	0.1775	0.1721	0.2027	0.1334
IQR.	0.2402	0.2305	0.2455	0.1789

H-25

component of e

	1	2	3	|e|
min.	-0.6466	-0.4884	-0.8159	0.0113
x 05	-0.2778	-0.2768	-0.3057	0.0940
Q1	-0.1202	-0.1028	-0.1175	0.1865
Q2	-0.0010	0.0039	-0.0037	0.2600
Q3	0.1118	0.1032	0.1132	0.3491
x 95	0.2794	0.2807	0.3311	0.5106
max.	0.4847	0.6409	0.8147	0.9188
MEAN	-0.0014	0.0028	-0.0023	0.2762
S.D.	0.1689	0.1647	0.1933	0.1291
IQR.	0.2320	0.2060	0.2307	0.1626

LAD

component of e

	1	2	3	|e|
min.	-0.5553	-0.5032	-0.6792	0.0182
x 05	-0.3010	-0.2900	-0.3554	0.0912
Q1	-0.1149	-0.1109	-0.1156	0.1814
Q2	-0.0023	0.0027	0.0081	0.2709
Q3	0.1140	0.1130	0.1103	0.3698
x 95	0.2804	0.2950	0.3381	0.5418
max.	0.4957	0.7282	0.8585	0.9295
MEAN	-0.0033	0.0042	0.0014	0.2862
S.D.	0.1772	0.1726	0.2031	0.1431
IQR.	0.2290	0.2239	0.2260	0.1884

TABLE

Estimation errors in Regressions (2.5.8), P.05
Errors, 901 Repetitions with Sample Size = 50

LSQ

	component of e			
	1	2	3	|e|
min.	-8.0041	-4.5403	-6.7121	0.0488
x 05	-0.7162	-0.6239	-0.6413	0.1474
Q1	-0.2229	-0.1956	-0.1896	0.3083
Q2	0.0136	0.0089	0.0117	0.4572
Q3	0.2263	0.2155	0.2351	0.7205
x 95	0.8704	0.7356	0.6929	2.0425
max.	8.8132	9.9729	16.8142	18.1879
MEAN	0.0350	0.0477	0.0641	0.7540
S.D.	0.8260	0.7210	1.0363	1.3094
IQR.	0.4493	0.4111	0.4247	0.4121

H-05

	component of e			
	1	2	3	|e|
min.	-0.5814	-0.5735	-0.9825	0.0272
x 05	-0.2904	-0.2867	-0.3300	0.0871
Q1	-0.1038	-0.1091	-0.1037	0.1908
Q2	-0.0021	0.0100	0.0131	0.2652
Q3	0.1113	0.1291	0.1278	0.3707
x 95	0.3031	0.2853	0.3449	0.5776
max.	0.7218	0.6441	0.6643	0.9872
MEAN	0.0031	0.0054	0.0115	0.2901
S.D.	0.1772	0.1777	0.2056	0.1456
IQR.	0.2151	0.2383	0.2314	0.1799

H-25

component of e

	1	2	3	|e|
min.	-0.5431	-0.5547	-1.0375	0.0193
x 05	-0.2754	-0.2806	-0.3188	0.0941
Q1	-0.1062	-0.1094	-0.1117	0.1836
Q2	-0.0020	0.0108	0.0086	0.2643
Q3	0.1081	0.1195	0.1274	0.3569
x 95	0.2879	0.2787	0.3311	0.5391
max.	0.6558	0.6299	0.7213	1.0407
MEAN	0.0020	0.0045	0.0097	0.2808
S.D.	0.1699	0.1699	0.1974	0.1337
IQR.	0.2143	0.2289	0.2391	0.1733

LAD

component of e

	1	2	3	|e|
min.	-0.6093	-0.6680	-1.1676	0.0278
x 05	-0.3130	-0.3163	-0.3538	0.1079
Q1	-0.1277	-0.1359	-0.1237	0.2128
Q2	-0.0108	0.0051	0.0107	0.3024
Q3	0.1291	0.1324	0.1429	0.4015
x 95	0.3269	0.3149	0.3629	0.5827
max.	0.7297	0.5780	0.7415	1.1772
MEAN	-0.0002	0.0035	0.0116	0.3183
S.D.	0.1948	0.1914	0.2183	0.1447
IQR.	0.2567	0.2683	0.2666	0.1887

TABLE

Estimation errors in Regressions (2.5.8), P.25
Errors, 901 Repetitions with Sample Size = 50

LSQ

	component of e			
	1	2	3	|e|
min.	-54.2074	-63.0177	-58.8688	0.1072
x 05	-3.4027	-3.1934	-3.4976	0.4217
Q1	-0.7078	-0.7449	-0.6495	0.9159
Q2	0.0770	-0.0411	0.0283	1.5095
Q3	0.7872	0.6588	0.5954	2.7550
x 95	3.2028	3.2246	2.6709	9.8776
max.	100.8236	84.7183	28.6847	127.6634
MEAN	0.0622	-0.1097	-0.3167	3.1701
S.D.	5.0252	4.7532	4.1705	7.4360
IQR.	1.4950	1.4037	1.2449	1.8391

H-05

	component of e			
	1	2	3	|e|
min.	-1.3053	-1.7315	-2.0755	0.0595
x 05	-0.7340	-0.7480	-0.8148	0.2315
Q1	-0.2760	-0.2905	-0.2779	0.4492
Q2	0.0046	-0.0163	0.0374	0.6420
Q3	0.2900	0.2766	0.3102	0.9034
x 95	0.7480	0.7016	0.7965	1.3515
max.	1.5147	1.4084	1.7282	2.3243
MEAN	0.0028	-0.0147	0.0247	0.7012
S.D.	0.4394	0.4393	0.4824	0.3570
IQR.	0.5660	0.5671	0.5881	0.4542

H-25

component of e

	1	2	3	|e|
min.	-1.3497	-1.7775	-2.1953	0.0582
x 05	-0.6483	-0.6749	-0.7720	0.1416
Q1	-0.2069	-0.2149	-0.2039	0.2995
Q2	0.0038	-0.0035	0.0270	0.4667
Q3	0.2096	0.2039	0.2410	0.7723
x 95	0.6437	0.6327	0.7613	1.3699
max.	1.6589	1.4440	1.9253	2.3826
MEAN	-0.0007	-0.0102	0.0247	0.5879
S.D.	0.3845	0.3926	0.4491	0.3980
IQR.	0.4165	0.4188	0.4449	0.4728

LAD

component of e

	1	2	3	|e|
min.	-1.3965	-1.7904	-2.5895	0.0433
x 05	-0.5961	-0.6402	-0.7663	0.1516
Q1	-0.1861	-0.2050	-0.1648	0.2695
Q2	-0.0065	-0.0091	0.0266	0.4057
Q3	0.1789	0.1705	0.2163	0.6646
x 95	0.5753	0.5819	0.7511	1.4334
max.	1.7746	1.6066	2.2541	2.8445
MEAN	-0.0022	-0.0125	0.0273	0.5531
S.D.	0.3742	0.3805	0.4479	0.4244
IQR.	0.3650	0.3755	0.3811	0.3951

TABLE

Estimation errors in Regressions (2.5.8), PARETO Errors, 901 Repetitions with Sample Size = 50

LSQ

component of e

	1	2	3	|e|
min.	-36.0003	-980.4569	-332.1250	0.1743
x 05	-3.6388	-3.7075	-3.0906	0.3951
Q1	-0.7377	-0.6827	-0.5823	0.8007
Q2	0.0351	-0.0395	-0.0617	1.4415
Q3	0.7075	0.6540	0.5604	2.9103
x 95	3.4494	3.2659	3.1309	11.3615
max.	3025.7315	74.5277	45.3557	3197.9140
MEAN	3.6978	-1.2811	-0.6510	7.1964
S.D.	101.0305	33.5422	13.1300	107.0914
IQR.	1.4452	1.3367	1.1428	2.1096

H-05

component of e

	1	2	3	|e|
min.	-0.8249	-0.8946	-0.9868	0.0354
x 05	-0.3780	-0.3440	-0.4158	0.1177
Q1	-0.1613	-0.1278	-0.1540	0.2323
Q2	-0.0062	0.0097	-0.0035	0.3328
Q3	0.1617	0.1489	0.1371	0.4617
x 95	0.4156	0.3710	0.4188	0.7438
max.	0.7454	0.7088	0.9334	1.1201
MEAN	0.0020	0.0123	-0.0060	0.3663
S.D.	0.2405	0.2160	0.2536	0.1862
IQR.	0.3230	0.2767	0.2911	0.2294

H-25

component of e

	1	2	3	|e|
min.	-0.6651	-0.6931	-0.8118	0.0228
x 05	-0.3173	-0.2970	-0.3528	0.0931
Q1	-0.1342	-0.1053	-0.1283	0.1997
Q2	-0.0034	0.0098	-0.0058	0.2870
Q3	0.1330	0.1297	0.1153	0.4054
x 95	0.3574	0.3063	0.3855	0.6301
max.	0.6647	0.7403	0.9387	0.9569
MEAN	0.0029	0.0104	-0.0041	0.3135
S.D.	0.2044	0.1827	0.2241	0.1648
IQR.	0.2672	0.2350	0.2436	0.2057

LAD

component of e

	1	2	3	|e|
min.	-0.5641	-0.6419	-0.8047	0.0160
x 05	-0.3145	-0.2867	-0.3872	0.0748
Q1	-0.1117	-0.1038	-0.1034	0.1669
Q2	0.0014	0.0013	-0.0010	0.2553
Q3	0.1146	0.1160	0.1094	0.3814
x 95	0.3300	0.2967	0.3659	0.6397
max.	0.6435	0.8389	1.0441	1.0518
MEAN	0.0023	0.0063	-0.0012	0.2929
S.D.	0.1918	0.1761	0.2205	0.1749
IQR.	0.2263	0.2199	0.2129	0.2145

TABLE

Estimation errors in Regressions (2.5.8), LOGISTIC Errors, 901 Repetitions with Sample Size = 50

LSQ

component of e

	1	2	3	|e|
min.	-0.9105	-0.9208	-0.8424	0.0116
x 05	-0.4154	-0.4093	-0.4513	0.1471
Q1	-0.1694	-0.1648	-0.1908	0.2824
Q2	0.0054	0.0054	-0.0160	0.4027
Q3	0.1675	0.1846	0.1611	0.5447
x 95	0.4429	0.4545	0.4781	0.7918
max.	0.8762	0.7674	0.9011	1.1891
MEAN	0.0010	0.0105	-0.0083	0.4241
S.D.	0.2626	0.2649	0.2790	0.1926
IQR.	0.3369	0.3494	0.3520	0.2623

H-05

component of e

	1	2	3	|e|
min.	-0.8811	-0.8542	-0.8829	0.0660
x 05	-0.4146	-0.4006	-0.4702	0.1500
Q1	-0.1662	-0.1677	-0.1824	0.2778
Q2	0.0011	0.0194	-0.0253	0.3864
Q3	0.1683	0.1668	0.1662	0.5263
x 95	0.3923	0.4516	0.4515	0.7582
max.	0.8594	0.7486	0.8808	1.1253
MEAN	-0.0010	0.0121	-0.0119	0.4128
S.D.	0.2568	0.2542	0.2726	0.1858
IQR.	0.3345	0.3345	0.3486	0.2484

H-25

component of e

	1	2	3	|e|
min.	-0.9307	-0.8587	-0.8388	0.0303
x 05	-0.4481	-0.4147	-0.4895	0.1524
Q1	-0.1684	-0.1553	-0.1920	0.2858
Q2	-0.0044	0.0127	-0.0268	0.3964
Q3	0.1841	0.1810	0.1763	0.5441
x 95	0.4219	0.4496	0.4598	0.7835
max.	0.9113	0.8057	0.8920	1.1184
MEAN	0.0012	0.0123	-0.0135	0.4248
S.D.	0.2640	0.2606	0.2824	0.1925
IQR.	0.3525	0.3363	0.3683	0.2583

LAD

component of e

	1	2	3	|e|
min.	-1.0060	-0.7728	-1.0496	0.0495
x 05	-0.4782	-0.4466	-0.5144	0.1626
Q1	-0.1780	-0.1781	-0.2191	0.3068
Q2	0.0021	0.0054	-0.0222	0.4482
Q3	0.1857	0.2013	0.1966	0.5995
x 95	0.4620	0.4802	0.5010	0.8481
max.	0.9533	0.8655	1.1225	1.3298
MEAN	-0.0012	0.0106	-0.0119	0.4645
S.D.	0.2867	0.2862	0.3108	0.2119
IQR.	0.3637	0.3794	0.4157	0.2927

TABLE

Estimation errors in Regressions (2.5.8), NORMAL Errors, 901 Repetitions with Sample Size = 100

LSQ

	component of e			
	1	2	3	|e|
min.	-0.2990	-0.3242	-0.2909	0.0148
x 05	-0.1587	-0.1749	-0.1744	0.0639
Q1	-0.0627	-0.0751	-0.0659	0.1130
Q2	0.0068	0.0028	0.0055	0.1595
Q3	0.0752	0.0695	0.0788	0.2084
x 95	0.1759	0.1655	0.1812	0.2853
max.	0.3306	0.3654	0.3428	0.4473
MEAN	0.0067	-0.0010	0.0041	0.1646
S.D.	0.1001	0.1024	0.1058	0.0680
IQR.	0.1379	0.1445	0.1447	0.0954

H-05

	component of e			
	1	2	3	|e|
min.	-0.2927	-0.3627	-0.3183	0.0231
x 05	-0.1610	-0.1740	-0.1808	0.0663
Q1	-0.0653	-0.0767	-0.0668	0.1174
Q2	0.0040	0.0020	0.0067	0.1635
Q3	0.0777	0.0701	0.0789	0.2138
x 95	0.1758	0.1654	0.1879	0.2867
max.	0.3559	0.3495	0.3200	0.4459
MEAN	0.0067	-0.0014	0.0047	0.1679
S.D.	0.1029	0.1041	0.1074	0.0692
IQR.	0.1430	0.1468	0.1457	0.0964

H-25

component of e

	1	2	3	|e|
min.	-0.3147	-0.3441	-0.3639	0.0149
x 05	-0.1696	-0.1811	-0.1764	0.0698
Q1	-0.0735	-0.0750	-0.0722	0.1217
Q2	0.0035	0.0006	0.0078	0.1709
Q3	0.0825	0.0716	0.0839	0.2257
x 95	0.1852	0.1852	0.1920	0.3075
max.	0.3907	0.3424	0.3271	0.4688
MEAN	0.0056	-0.0016	0.0053	0.1773
S.D.	0.1094	0.1099	0.1133	0.0740
IQR.	0.1560	0.1465	0.1561	0.1040

LAD

component of e

	1	2	3	|e|
min.	-0.3941	-0.3781	-0.4627	0.0232
x 05	-0.2147	-0.2178	-0.2104	0.0745
Q1	-0.0802	-0.0882	-0.0808	0.1436
Q2	0.0061	0.0051	0.0082	0.1967
Q3	0.0841	0.0846	0.0911	0.2674
x 95	0.2070	0.2081	0.2257	0.3686
max.	0.4514	0.4340	0.4142	0.5279
MEAN	0.0039	-0.0011	0.0064	0.2067
S.D.	0.1275	0.1294	0.1330	0.0893
IQR.	0.1642	0.1728	0.1719	0.1238

TABLE

Estimation errors in Regressions (2.5.8), H.05
Errors, 901 Repetitions with Sample Size = 100

LSQ

component of e

	1	2	3	|e|
min.	-0.3662	-0.2869	-0.3584	0.0144
x 05	-0.1647	-0.1658	-0.1751	0.0620
Q1	-0.0684	-0.0596	-0.0715	0.1116
Q2	-0.0033	0.0053	-0.0029	0.1551
Q3	0.0674	0.0730	0.0694	0.2052
x 95	0.1609	0.1652	0.1746	0.2837
max.	0.3134	0.3335	0.3170	0.4325
MEAN	-0.0017	0.0043	-0.0019	0.1623
S.D.	0.1013	0.0986	0.1054	0.0688
IQR.	0.1359	0.1326	0.1409	0.0936

H-05

component of e

	1	2	3	|e|
min.	-0.3179	-0.2797	-0.3363	0.0124
x 05	-0.1639	-0.1524	-0.1703	0.0575
Q1	-0.0684	-0.0581	-0.0697	0.1067
Q2	-0.0030	0.0032	-0.0014	0.1531
Q3	0.0587	0.0708	0.0636	0.1973
x 95	0.1526	0.1661	0.1710	0.2707
max.	0.3133	0.3380	0.3357	0.4433
MEAN	-0.0027	0.0040	-0.0027	0.1554
S.D.	0.0959	0.0953	0.1016	0.0666
IQR.	0.1271	0.1290	0.1333	0.0906

H-25

component of e

	1	2	3	|e|
min.	-0.3721	-0.3028	-0.3789	0.0235
x 05	-0.1890	-0.1794	-0.1939	0.0677
Q1	-0.0787	-0.0736	-0.0832	0.1292
Q2	-0.0021	0.0008	-0.0032	0.1772
Q3	0.0714	0.0859	0.0742	0.2327
x 95	0.1856	0.1959	0.1950	0.3153
max.	0.3341	0.3424	0.3880	0.4586
MEAN	-0.0030	0.0040	-0.0037	0.1827
S.D.	0.1121	0.1123	0.1173	0.0746
IQR.	0.1501	0.1595	0.1574	0.1034

LAD

component of e

	1	2	3	|e|
min.	-0.4047	-0.3734	-0.4012	0.0274
x 05	-0.2245	-0.2153	-0.2249	0.0823
Q1	-0.0901	-0.0882	-0.0970	0.1444
Q2	-0.0019	-0.0001	0.0022	0.2024
Q3	0.0895	0.0951	0.0833	0.2711
x 95	0.2163	0.2253	0.2285	0.3661
max.	0.3707	0.3621	0.4746	0.5494
MEAN	-0.0026	0.0031	-0.0021	0.2130
S.D.	0.1318	0.1320	0.1371	0.0904
IQR.	0.1796	0.1833	0.1804	0.1266

TABLE

Estimation errors in Regressions (2.5.8), H.25 Errors, 901 Repetitions with Sample Size = 100

LSQ

component of e

	1	2	3	|e|
min.	-0.4633	-0.5067	-0.5603	0.0183
x 05	-0.2281	-0.2095	-0.2487	0.0727
Q1	-0.0839	-0.0854	-0.0849	0.1432
Q2	0.0064	0.0151	0.0031	0.2061
Q3	0.0883	0.0962	0.0967	0.2802
x 95	0.2249	0.2183	0.2619	0.4124
max.	0.4365	0.4708	0.4800	0.6443
MEAN	0.0042	0.0073	0.0020	0.2188
S.D.	0.1362	0.1329	0.1488	0.1025
IQR.	0.1723	0.1816	0.1816	0.1371

H-05

component of e

	1	2	3	|e|
min.	-0.2858	-0.2974	-0.4021	0.0119
x 05	-0.1596	-0.1538	-0.1802	0.0569
Q1	-0.0569	-0.0588	-0.0658	0.1028
Q2	0.0023	0.0107	-0.0002	0.1478
Q3	0.0626	0.0718	0.0631	0.1997
x 95	0.1644	0.1486	0.1698	0.2864
max.	0.2963	0.3250	0.3419	0.4546
MEAN	0.0027	0.0052	-0.0010	0.1561
S.D.	0.0965	0.0925	0.1076	0.0712
IQR.	0.1196	0.1306	0.1289	0.0969

H-25

component of e

	1	2	3	|e|
min.	-0.2744	-0.2644	-0.3411	0.0164
x 05	-0.1342	-0.1277	-0.1531	0.0469
Q1	-0.0507	-0.0506	-0.0520	0.0852
Q2	0.0034	0.0039	-0.0007	0.1206
Q3	0.0517	0.0568	0.0516	0.1633
x 95	0.1387	0.1269	0.1446	0.2411
max.	0.2416	0.2553	0.2583	0.3642
MEAN	0.0021	0.0027	-0.0012	0.1293
S.D.	0.0798	0.0776	0.0891	0.0601
IQR.	0.1024	0.1074	0.1036	0.0780

LAD

component of e

	1	2	3	|e|
min.	-0.2782	-0.3340	-0.3654	0.0043
x 05	-0.1255	-0.1332	-0.1572	0.0412
Q1	-0.0486	-0.0500	-0.0541	0.0766
Q2	0.0004	0.0026	0.0010	0.1149
Q3	0.0548	0.0524	0.0518	0.1679
x 95	0.1428	0.1311	0.1433	0.2512
max.	0.2429	0.2726	0.3440	0.4101
MEAN	0.0034	0.0001	-0.0021	0.1278
S.D.	0.0812	0.0801	0.0884	0.0669
IQR.	0.1034	0.1024	0.1059	0.0913

TABLE

Estimation errors in Regressions (2.5.8), DEXP Errors, 901 Repetitions with Sample Size = 100

LSQ

	component of e			
	1	2	3	|e|
min.	-0.5417	-0.4601	-0.5786	0.0162
x 05	-0.2250	-0.2301	-0.2333	0.0804
Q1	-0.0895	-0.1039	-0.0907	0.1560
Q2	0.0002	-0.0023	-0.0029	0.2198
Q3	0.0921	0.0989	0.0933	0.2924
x 95	0.2278	0.2345	0.2880	0.4271
max.	0.4057	0.4309	0.7925	0.8007
MEAN	-0.0008	-0.0015	0.0072	0.2311
S.D.	0.1398	0.1465	0.1544	0.1070
IQR.	0.1816	0.2028	0.1840	0.1364

H-05

	component of e			
	1	2	3	|e|
min.	-0.4604	-0.3997	-0.4993	0.0223
x 05	-0.1856	-0.2020	-0.2036	0.0793
Q1	-0.0801	-0.0853	-0.0832	0.1361
Q2	0.0003	0.0009	-0.0044	0.1879
Q3	0.0831	0.0862	0.0815	0.2531
x 95	0.2037	0.2055	0.2424	0.3646
max.	0.3416	0.4007	0.5542	0.5631
MEAN	0.0025	0.0016	0.0045	0.2011
S.D.	0.1205	0.1266	0.1334	0.0888
IQR.	0.1632	0.1715	0.1647	0.1170

H-25

component of e

	1	2	3	|e|
min.	-0.3789	-0.3804	-0.4606	0.0169
x 05	-0.1798	-0.1834	-0.1971	0.0681
Q1	-0.0764	-0.0802	-0.0794	0.1314
Q2	0.0047	0.0018	-0.0059	0.1820
Q3	0.0821	0.0817	0.0754	0.2395
x 95	0.1917	0.1986	0.2316	0.3374
max.	0.3377	0.3816	0.4338	0.5506
MEAN	0.0038	0.0028	0.0027	0.1907
S.D.	0.1149	0.1191	0.1269	0.0842
IQR.	0.1584	0.1619	0.1548	0.1081

LAD

component of e

	1	2	3	|e|
min.	-0.3787	-0.3865	-0.4246	0.0139
x 05	-0.1806	-0.1879	-0.2037	0.0579
Q1	-0.0742	-0.0719	-0.0739	0.1246
Q2	0.0040	0.0023	-0.0018	0.1733
Q3	0.0818	0.0815	0.0744	0.2342
x 95	0.1903	0.1867	0.2179	0.3522
max.	0.3905	0.3873	0.5161	0.6515
MEAN	0.0029	0.0027	0.0018	0.1847
S.D.	0.1135	0.1153	0.1254	0.0882
IQR.	0.1560	0.1535	0.1483	0.1097

TABLE

Estimation errors in Regressions (2.5.8), P.05 Errors, 901 Repetitions with Sample Size = 100

LSQ

	component of e			
	1	2	3	|e|
min.	-31.4961	-5.7912	-53.9663	0.0379
x 05	-0.7031	-0.6525	-0.5469	0.1112
Q1	-0.1836	-0.1689	-0.1805	0.2435
Q2	-0.0080	0.0019	-0.0111	0.3790
Q3	0.1812	0.1662	0.1522	0.6019
x 95	0.6897	0.7706	0.6661	2.3726
max.	18.5581	16.5248	17.8837	62.5396
MEAN	-0.0478	0.0689	-0.0320	0.7783
S.D.	1.5111	1.1389	2.0215	2.6586
IQR.	0.3647	0.3351	0.3326	0.3584

H-05

	component of e			
	1	2	3	|e|
min.	0.3633	-0.4446	-0.4138	0.0133
x 05	-0.1876	-0.1977	-0.2275	0.0701
Q1	-0.0823	-0.0770	-0.0854	0.1273
Q2	-0.0013	-0.0034	- 0.0001	0.1823
Q3	0.0735	0.0755	0.0910	0.2465
x 95	0.1967	0.1861	0.2150	0.3661
max.	0.3615	0.3260	0.6069	0.6856
MEAN	-0.0016	-0.0032	-0.0001	0.1942
S.D.	0.1174	0.1162	0.1368	0.0907
IQR.	0.1558	0.1524	0.1764	0.1192

H-25

component of e

	1	2	3	$\|e\|$
min.	-0.3556	-0.4228	-0.3958	0.0275
x 05	-0.1842	-0.1961	-0.2166	0.0750
Q1	-0.0804	-0.0772	-0.0841	0.1273
Q2	-0.0022	0.0016	0.0035	0.1812
Q3	0.0777	0.0755	0.0901	0.2379
x 95	0.1882	0.1800	0.2102	0.3526
max.	0.3816	0.3038	0.5940	0.6665
MEAN	-0.0002	-0.0037	0.0005	0.1904
S.D.	0.1159	0.1127	0.1320	0.0852
IQR.	0.1181	0.1527	0.1742	0.1106

LAD

component of e

	1	2	3	$\|e\|$
min.	-0.4300	-0.5381	-0.5253	0.0249
x 05	-0.2074	-0.2139	-0.2433	0.0831
Q1	-0.0886	-0.0931	-0.0886	0.1480
Q2	-0.0015	-0.0042	0.0097	0.2072
Q3	0.0915	0.0918	0.1042	0.2732
x 95	0.2211	0.2125	0.2371	0.3943
max.	0.3909	0.3241	0.6375	0.6758
MEAN	0.0022	-0.0027	0.0040	0.2176
S.D.	0.1323	0.1308	0.1481	0.0958
IQR.	0.1801	0.1849	0.1928	0.1252

TABLE

Estimation errors in Regressions (2.5.8), P.25
Errors, 901 Repetitions with Sample Size = 100

LSQ

	component of e			
	1	2	3	|e|
min.	-54.0462	-153.6104	-101.1865	0.0940
x 05	-2.4966	-2.4923	-2.4738	0.3800
Q1	-0.6525	-0.5430	-0.5664	0.7769
Q2	0.0179	0.0112	-0.0329	1.2603
Q3	0.6717	0.6468	0.4846	2.3939
x 95	3.1150	2.7553	2.2266	8.2693
max.	342.6049	55.4104	85.7701	388.8612
MEAN	0.5407	-0.0951	-0.1113	3.4277
S.D.	12.4781	6.9587	5.9398	15.0983
IQR.	1.3243	1.1898	1.0510	1.6170

H-05

	component of e			
	1	2	3	|e|
min.	-1.1501	-1.0182	-1.1599	0.0346
x 05	-0.4841	-0.4712	-0.5044	0.1470
Q1	-0.1799	-0.1749	-0.1991	0.2921
Q2	0.0069	-0.0065	-0.0047	0.4353
Q3	0.1866	0.1897	0.1676	0.6055
x 95	0.4786	0.4986	0.5214	0.8832
max.	0.9995	1.0064	1.3364	1.3781
MEAN	0.0012	0.0055	-0.0031	0.4617
S.D.	0.2872	0.2896	0.3110	0.2231
IQR.	0.3665	0.3646	0.3668	0.3134

H-25

component of e

	1	2	3	|e|
min.	-1.0286	-0.8373	-1.1197	0.0095
x 05	-0.3489	-0.3256	-0.4161	0.0887
Q1	-0.1238	-0.1159	-0.1420	0.1868
Q2	0.0068	-0.0021	0.0019	0.2838
Q3	0.1267	0.1250	0.1167	0.4279
x 95	0.3361	0.3437	0.4616	0.7857
max.	0.9810	1.0468	1.4061	1.4394
MEAN	0.0009	0.0041	0.0011	0.3340
S.D.	0.2107	0.2143	0.2567	0.2111
IQR.	0.2505	0.2409	0.2587	0.2411

LAD

component of e

	1	2	3	|e|
min.	-0.9605	-0.6871	-1.1030	0.0236
x 05	-0.2931	-0.2883	-0.3761	0.0948
Q1	-0.1189	-0.1120	-0.1275	0.1890
Q2	0.0093	0.0018	0.0045	0.2666
Q3	0.1146	0.1138	0.1136	0.3658
x 95	0.3181	0.2927	0.4290	0.6799
max.	0.9926	1.0180	1.4172	1.4595
MEAN	0.0018	0.0020	0.0002	0.3066
S.D.	0.1940	0.1900	0.2380	0.1905
IQR.	0.2335	0.2258	0.2411	0.1767

TABLE

Estimation errors in Regressions (2.5.8), PARETO Errors, 901 Repetitions with Sample Size = 100

LSQ

	component of e			
	1	2	3	|e|
min.	-175.4391	-112.4396	-103.1157	0.1255
x 05	-2.8054	-2.5749	-2.5706	0.3725
Q1	-0.6839	-0.5959	-0.5352	0.7954
Q2	-0.0084	-0.0092	-0.0471	1.3262
Q3	0.6986	0.5667	0.4757	2.2638
x 95	2.9148	2.2901	2.2333	8.3724
max.	107.9225	165.1103	24.7446	241.3110
MEAN	0.1222	-0.1340	-0.2444	3.3379
S.D.	9.1225	8.7747	4.9603	13.1817
IQR.	1.3825	1.1626	1.0109	1.4684

H-05

	component of e			
	1	2	3	|e|
min.	-0.4240	-0.4855	-0.6793	0.0075
x 05	-0.2496	-0.2604	-0.2803	0.0827
Q1	-0.1113	-0.1068	-0.1148	0.1652
Q2	-0.0063	-0.0101	-0.0116	0.2332
Q3	0.0961	0.0883	0.0951	0.3150
x 95	0.2444	0.2377	0.2763	0.4656
max.	0.4965	0.5024	0.5699	0.6810
MEAN	-0.0045	-0.0092	-0.0083	0.2481
S.D.	0.1513	0.1493	0.1716	0.1146
IQR.	0.2074	0.1950	0.2100	0.1498

H-25

component of e

	1	2	3	|e|
min.	-0.3798	-0.4149	-0.5931	0.0198
x 05	-0.2202	-0.2123	-0.2370	0.0693
Q1	-0.0880	-0.0928	-0.0938	0.1380
Q2	-0.0047	-0.0121	-0.0080	0.1965
Q3	0.0846	0.0712	0.0835	0.2663
x 95	0.2047	0.2013	0.2418	0.3880
max.	0.3920	0.4341	0.5232	0.6295
MEAN	-0.0047	-0.0080	-0.0063	0.2085
S.D.	0.1277	0.1244	0.1463	0.0989
IQR.	0.1726	0.1640	0.1772	0.1283

LAD

component of e

	1	2	3	|e|
min.	-0.4316	-0.3657	-0.5865	0.0158
x 05	-0.1836	-0.1715	-0.2107	0.0559
Q1	-0.0731	-0.0733	-0.0795	0.1113
Q2	-0.0034	-0.0076	-0.0055	0.1629
Q3	0.0681	0.0604	0.0739	0.2323
x 95	0.1737	0.1713	0.2279	0.3627
max.	0.3949	0.4002	0.5558	0.5894
MEAN	-0.0028	-0.0052	-0.0023	0.1806
S.D.	0.1111	0.1077	0.1344	0.0969
IQR.	0.1413	0.1337	0.1534	0.1210

TABLE

Estimation errors in Regressions (2.5.8), LOGISTIC Errors, 901 Repetitions with Sample Size = 100

LSQ

	component of e			
	1	2	3	|e|
min.	-0.5957	-0.5831	-0.8284	0.0125
x 05	-0.3028	-0.3038	-0.3128	0.1057
Q1	-0.1112	-0.1269	-0.0985	0.1938
Q2	0.0058	0.0043	0.0132	0.2742
Q3	0.1165	0.1279	0.1291	0.3680
x 95	0.2885	0.3137	0.3140	0.5325
max.	0.6296	0.6822	0.6482	0.8480
MEAN	-0.0004	0.0064	0.0122	0.2905
S.D.	0.1792	0.1869	0.1859	0.1315
IQR.	0.2277	0.2548	0.2276	0.1742

H-05

	component of e			
	1	2	3	|e|
min.	-0.6441	-0.5894	-0.7582	0.0139
x 05	-0.2926	-0.2968	-0.2956	0.0942
Q1	-0.1155	-0.1138	-0.1015	0.1897
Q2	0.0028	0.0019	0.0093	0.2694
Q3	0.1097	0.1369	0.1224	0.3578
x 95	0.3033	0.2997	0.3054	0.5106
max.	0.6220	0.7329	0.7156	0.8141
MEAN	0.0005	0.0084	0.0122	0.2831
S.D.	0.1735	0.1836	0.1813	0.1290
IQR.	0.2252	0.2507	0.2239	0.1681

H-25

component of e

	1	2	3	|e|
min.	-0.7217	-0.5535	-0.7542	0.0399
x 05	-0.2934	-0.2999	-0.2885	0.1002
Q1	-0.1230	-0.1189	-0.1104	0.2015
Q2	0.0049	0.0041	0.0143	0.2767
Q3	0.1144	0.1477	0.1269	0.3731
x 95	0.3054	0.3098	0.3220	0.5258
max.	0.6061	0.7618	0.7706	0.8468
MEAN	0.0007	0.0104	0.0116	0.2947
S.D.	0.1789	0.1892	0.1907	0.1323
IQR.	0.2374	0.2665	0.2373	0.1716

LAD

component of e

	1	2	3	|e|
min.	-0.7337	-0.7656	-0.7814	0.0093
x 05	-0.3295	-0.3276	-0.3373	0.1251
Q1	-0.1246	-0.1341	-0.1255	0.2193
Q2	0.0032	0.0163	0.0098	0.3079
Q3	0.1447	0.1457	0.1493	0.4085
x 95	0.3351	0.3424	0.3462	0.5888
max.	0.6019	0.8494	0.8793	0.9452
MEAN	0.0061	0.0112	0.0109	0.3260
S.D.	0.1989	0.2053	0.2127	0.1442
IQR.	0.2693	0.2798	0.2748	0.1892

BIBLIOGRAPHY

Abdelmalek, N.N. Linear L_1 Approximation for a Discrete Point Set and L_1 Solutions of Overdetermined Equations. *JACM*, 1971, *18*, 41–47.

Abdelmalek, N.N. On the Discrete Linear L_1 Approximation and L_1 Solutions of Overdetermined Equations. *J. Approx. Theory*, 1974, *11*, 38–53.

Abdelmalek, N.N. An Efficient Method for the Discrete Linear L_1 Approximation Problem. *Math. of Computation*, 1975, *29*, 844–850.

Amemiya, T. *The Two Stage Least Absolute Deviations* Estimators. Technical Report 297, Institute for Mathematical Studies in the Social Sciences, Stanford University, 1979.

An, Hong-Zhi And Chen, Zhao-Guo. On Convergence of LAD Esitmates in Autoregression with Infinite Variance. *J. Multivariate Analysis*, 1982, *12*, 335–345.

Anderson, D. and Steiger, W.L. *A Comparison of Methods for Discrete L_1 Curve Fitting*. Technical Report 96, Department of Computer Science, Rutgers University, 1982.

Anderssen, R.S., Bloomfield, P., and McNeil, D. *Spline Functions in Data Analysis*. Technical Report 69, Series 2, Department of Statistics, Princeton University, 1974.

Andrews, D.F. A Robust Method for Multiple Linear Regression. *Technometrics*, 1974, *16*, 523–532.

Appa, G. and Smith, C. On L_1 and Chebyshev Estimation. *J. Math. Prog.*, 1973, *5*, 78–87.

Armstrong, R.D., Elam, J.J. and Hultz, J.W. Obtaining Least Absolute Value Estimates for a Two-Way Classification Model. *Commun. Statist. B*, 1977, *6*, 365–381.

Armstrong, R.D. and Frome, E.L. A Comparison of Two Algorithms for Absolute Deviation Curve Fitting. *J. Amer. Statist. Assoc.*, 1976, *71*, 328–330.

Armstrong, R.D. and Frome, E.L. A Special Purpose Linear Programming Algorithm for Obtaining Least Absolute Value Estimates in a Linear Model with Dummy Variables. *Commun. Statist. B,* 1977, *6,* 383–398.

Armstrong, R.D. and Frome, E.L. Least-Absolute-Value Estimators for One-Way and Two-Way Tables. *Naval Research Logistics Quarterly,* 1979, *22,* 79–96.

Armstrong, R.D., Frome, E.L. and Kung, D.S. A Revised Simplex Algorithm for the Absolute Deviation Curve Fitting Problem. *Commun. Statist. B,* 1979, *8,* 175–190.

Armstrong, R.D. and Godfrey, J. Two Linear Programming Algorithms for the Linear Discrete L_1 Norm Problem. *Math. of Computation,* 1979, *33,* 289–300.

Armstrong, R.D. and Hultz, J.W. An Algorithm for a Restricted Discrete Approximation Problem in the L_1 Norm. *SIAM J. Numer. Anal.,* 1977, *14,* 555–565.

Ashar, V.G. and Wallace, T.D. A Sampling Study of Minimum Absolute Deviation Estimation. *Operations Research,* 1963, *11,* 747–758.

Avis, D. and Chvatal, V. Notes on Bland's Pivoting Rule. *Mathematical Programming Study,* 1978, *8,* 24–34.

Barrodale, I. and Roberts, F.D.K. Applications of Mathematical Programming to L_p Approximation. In Rosen, J.B., Mangasarian, O.L., and Ritter, K. (Eds.), *Nonlinear Programming,* New York: Academic Press, 1970.

Barrodale, I. and Roberts, F.D.K. *An Improved Algorithm For Discrete L_1 Linear Approximation.* Technical Report tsr 1172, Math. Research Center, U. of Wisconsin, 1972.

Barrodale, I. and Roberts, F.D.K. An Improved Algorithm for Discrete L_1 Linear Approximation. *SIAM J. Numer. Anal.* 1973, *10,* 839–848.

Barrodale, I. and Roberts, F.D.K. Algorithm 478: Solution of an Overdetermined System of Equations in the L_1 Norm. *CACM,* 1974, *14,* 319–320.

Barrodale, I. and Roberts, F.D.K. Algorithms for Restricted Least Absolute Value Estimation. *Comm. Statist B*, 1977, *6*, 353-363.

Barrodale, I. and Roberts, F.D.K. An Efficient Algorithm for Discrete L_1 LInear Approximation with Linear Constraints. *SIAM J. Numer. Anal.*, 1978, *15*, 603-611.

Barrodale, I. and Young, A. Algorithms for Best L_1 and L_∞ Linear Approximations on a Discrete Set. *Numer. Math.*, 1966, *8*, 295-306.

Bartels, R.H. A Penalty Linear Programming Method Using Reduced Gradient Basis-Exchange Techniques. *Linear Algebra and its Applications*, 1980, *29*, 17-32.

Bartels, R.H. and Conn, A.R. Least Absolute Regression: A Special Case of Piecewise Linear Minimization. *Comm. Statist. B*, 1977, *6*, 329-339.

Bartels, R.H. and Conn, A.R. Linearly Constrained Discrete L_1 Problems. *ACM Trans. Math. Software*, 1978, *6*, 594-608.

Bartels, R.H., Conn, A.R. and Sinclair, J. W. Minimization Techniques for Piecewise Differentiable ,Functions: the L_1 Solution to an Overdetermined ,System. *SIAM J. Numer. Anal.*, 1978, *15*, 224-241.

Bassett, G. and Koenker, R. Asymptotic Theory of Least Absolute Error Regression. *J. Amer. Statist. Assoc.*, 1978, *73*, 618-622.

Beaton, A.E. and Tukey, J.W. The Fitting of Power Series, Meaning Polynomials, Illustrated on Band-Spectographic Data. *Technometrics*, 1974, *16*, 147-185.

Belsley, D.A., Kuh, E., and Welsch, R.E. *Regression Diagnostics: Identifying Influential Data and Sources of Collinearity*. New York: John Wiley 1980.

Ben-Israel, A, and Charnes, A. An Explicit Solution of a Special Class of Linear Programming Problems. *Operations Research*, 1968, *16*, 1166-1175.

Billingsley, P. The Lindeberg-Levy Theorem for Martingales. *Proc. Amer. Math. Soc.*, 1961, *12*, 788-792.

Billingsley, P. *Convergence of Probability Measures.* New York: John Wiley 1968.

Bloomfield, P. *Least Absolute Deviations Regression.* SINAPE, 1982.

Bloomfield, P. and Steiger, W.L. Least Absolute Deviations Curve-Fitting. *SIAM J. Scientific and Statistical* Computing, 1980a, *1*, 290-301.

Bloomfield, P. and Steiger, W.L. Letter to the Editor. Technometrics, 1980b, *22*, 450.

Chambers, J.M. Algorithm 410: Partial Sorting. *CACM*, 1971, *14*, 357-358.

Charnes, A., Cooper, W.W., and Ferguson, R.O. Optimal Estimation of Executive Compensation by Linear Programming. *Mgt. Sci.*, 1955, *1*, 138-151.

Chvatal, V. *Linear Programming.* San Francisco, Cal.: W. H. Freeman 1983.

Claerbout, J.F. and Muir, F. Robust Modeling with Erratic Data. *Geophysics*, 1973, *38*, 826-844.

Clark, D.I. *Finite Algorithms for Linear Optimisation Problems.* PhD thesis, The Australian National University, 1981.

Clark, D.I. The Mathematical Structure of Huber's M-Estimator. *SIAM J. Scientific and Statistical Computing*, to appear, , .

Conn, A.R. Linear Programming Via a Non-Differentiable Penalty Function. *SIAM J. Numer. Anal.*, 1976, *13*, 145-154.

Cook, R.D. and Weisberg, S. Characterizations of an Empirical Influence Function for Detecting Influential Cases in Regression. *Technometrics*, 1980, *22*, 495-508.

Cook, R.D. and Weisberg, S. *Residuals and Influence in Regression.* New York: Chapman and Hall 1982.

Cox, D.D. Asymptotics for M-type Smoothing Splines. *Ann. Statist.*, 1983, *11*, 530-551.

Craven, P. and Wahba, G. Smoothing Noisy Data with Spline Functions: Estimating the Correct Degree of Smoothing by the Method of Generalized Cross-validation. *Numer. Math.*, 1979, *31*, 377–403.

Daniel, C. and Wood, F.S. *Fitting Equations to Data: Computer Analysis of Multifactor Data.* New York: John Wiley 1980. (2nd. edition).

Dantzig, G.B. Programming of Interdependent Activities, II. *Econometrica*, 1949, *17*, 200–211.

Dantzig, G.B. *Expected Number of Steps of the Simplex Method for a Linear Program with Convexity* Constraint. Technical Report 80–3, Systems Optimization Laboratory, 1980.

Dantzig, G.B. Khachian's Algorithm: A Comment. *SIAM News*, 1980, *13(5)*, 1.

Davies, M. Linear Approximation Using the Criterion of Least Total Deviation. *J. Roy. Statist. Soc. B*, 1965, *29*, 101–109.

De Boor, C. *A Practical Guide to Splines.* New York: Springer-Verlag 1978.

Donoho, D.L. Breakdown Properties of Multivariate Location Estimators. Ph.D. qualifying paper, Department of Statistics, Harvard University.

Dunham, J.R., Kelly, D.G., and Tolle, J.W. *Some Experimental Results Concerning the Expected Number of Pivots for Solving Randomly Generated Linear Programs.* Technical Report 77–16, Operations Research and Systems Analysis Department, University of North Carolina, 1977.

Dutter, R. *Robust Regression: Different Approaches to Numerical Solutions and Algorithms.* Technical Report 6, Fachgruppe f. Statist., Federal Institute of Technology, 1975.

Dutter, R. Algorithms for the Huber Estimator in Multiple Regression. *Computing*, 1977, *18*, 167–176.

Dutter, R. and Huber, P.J. Numerical Methods for the Nonlinear Robust Regression Problem. *J. Statist. Comp. Simul.*, 1981, *13*, 79–113.

Edgeworth, F.Y. A New Method of Reducing Observations Relating to Several Quantities. *Phil. Mag. (Fifth Series)*, 1887, *24*, 222–223.

Edgeworth, F.Y. On A New Method of Reducing Observations Relating to Several Quantities. *Phil. Mag. (Fifth Series)*, 1888, *25*, 184–191.

Eisenhart, C. Boscovitch and the Combination of Observations. In Whyte, L.L. (Ed.), *Roger Joseph Boscovitch*, New York: Fordham University Press, 1961.

Emerson, J.D. and Hoaglin, D.C. Analysis of Two-Way Tables by Medians. In Hoaglin, D.C, Mosteller, F., and Tukey, J. (Eds.), *Understanding Robust and Exploratory Data Analysis*, New York: John Wiley, 1983.

Fair, R.C. On the Robust Estimation of Econometric Models. *Ann. Econ. Soc. Measurement*, 1974, *3*, 667–677.

Fair, R.C. and Peck, J.K. *A Note on an Iterative Technique for Absolute Deviations Curve Fitting*. Technical Report, Department of Economics, Yale University, 1974.

Feller, W. *An Introduction to Probability Theory and Its Applications, Volume II*. New York: John Wiley 1971.

Fisher, L. and McDonald, J. *Fixed Effects Analysis of Variance.* New York: Academic Press 1978.

Fisher, W.D. A Note on Curve Fitting with Minimum Deviations by Linear Programming. *J. Amer. Statist. Assoc.*, 1961, *56*, 359–362.

Gacs, P. and Lovasz, L. Khachiyan's Algorithm for Linear Programming. *Linear Programming Study*, 1981, *14*, 61–68.

Gallant, A.R. and Gerig, T.M. *Comments on Computing Minimum Absolute Deviations Regressions by Iterative Least Squares Regressions and by Linear Programming*. Technical Report 911, Institute of Statistics, North Carolina State University, 1974.

Gentle, J.E. Least Absolute Values Estimation: An Introduction. *Commun. Statist. B*, 1977, *6*, 313–328.

Gentle, J.E. and Hanson, T.A. *Variable Selection Under L_1*. American Statistical Association, Washington, D.C., 1977.

Gentle, J.E., Kennedy, W.J., and Sposito, V.A. On Least Absolute Deviations Estimations. *Commun. Statist. A*, 1977, *6*, 839–845.

Gilsinn, J., Hoffman, K., Jackson, R.H.F., Leyendecker, E., Saunders, P., and Shier, D. Methodology and Analysis for Comparing Discrete Linear L_1 Approximation Codes. *Commun. Statist. B*, 1977, *6*, 399–413.

Goldfarb, D. and Reid, J.K. A Practicable Steepest-Edge Simplex Algorithm. *Math. Programming*, 1977, *12*, 361–371.

Goodall, C. Examining Residuals. In Hoaglin, D.C., Mosteller, F., and Tukey, J. (Eds.), *Understanding Robust and Exploratory Data Analysis*, : John Wiley, 1983.

Gross, S. and Steiger, W.L. Least Absolute Deviation Estimates in Autoregression with Infinite Variance. *J. Applied Probability*, 1979, *16*, 104–116.

Grotschel, M., Lovasz, L., and Schrijver, A. The Ellipsoidal Method and Its Consequences in Combinatorial Optimization. *Combinatorica*, 1981, *1*, 169–197.

Hampel, F.R. The Influence Curve and its Role in Robust Estimation. *J. Amer. Statist. Assoc.*, 1974, *62*, 1179–1186.

Harris, T. Regression Using Minimum Absolute Deviations. *American Statistician*, 1950, *4*, 14–15.

Hill, R.W. and Holland, P.W. Two Robust Alternatives to Least-Squares Regression. *J. Amer. Statist. Assoc.*, 1977, *72*, 828–833.

Hogg, R.V. An Introduction to Robust Estimation. In Launer, R.L. and Wilkinson, G.N. (Eds.), *Robustness in Statistics*, New York: Academic Press, 1979.

Holland, P.W. and Welsch, R.E. Robust Regression Using Iteratively Reweighted Least Squares. *Commun. Statist. a,* 1977, *6,* 813-827.

Huber, P.J. Robust Regression, Asymptotics, Conjectures and Monte-Carlo. *Ann. Statist.,* 1973, *1,* 799-821.

Huber, P.J. *SIAM Regional Conference Series in Applied Mathematics.* Volume 27: *Robust Statistical Procedures.* Philadelphia, Pa.: Society for Industrial and Applied Mathematics 1977.

Huber, P.J. Robust Smoothing. In Launer, R.L. and Wilkinson, G.N. (Eds.), *Robustness in Statistics,* New York: Academic Press, 1979.

Huber, P.J. *Robust Statistics.* New York: John Wiley 1981.

Ibragimov, I.A. and Linnik, Y. V. *Independent and Stationary Sequences of Random Variables.* Gronigen: Wolters-Noordhoff 1971.

Jaeckel, L. Estimating Regression Coefficients by Minimizing the Dispersion of the Residuals. *Ann. Math. Statist.,* 1972, *43,* 1449-1458.

Jeroslow, R.G. The Simplex Algorithm with Pivot Rule of Maximizing Criterion Improvement. *Discrete Math.,* 1973, *4,* 363-377.

Johnson, N.L. and Leone, F.C. *Statistics and Experimental Design.* New York: John Wiley 1964.

Kanter, M. and Hannan, E.J. Autoregressive Processes with Infinite Variance. *J. Applied Prob.,* 1977, *14,* 411-415.

Kanter, M. and Steiger, W.L. Regression and Autoregression with Infinite Variance. *Adv. Applied Prob.,* 1974, *6,* 768-783.

Kanter, M. and Steiger, W.L. *Estimating Linear Relationships for Models Based on Random Variables with Infinite Variance.* The Center of Mathematical Statistics of the Ministry of Education, Rumania, Bucharest, Rumania, 1977. (The Brasov conference).

Kantorovich, L.V. Mathematical Methods in the Organization and Planning of Production. *Management Sci.*, 1960, *6*, 366–422. (translated from a 1939 paper, in Russian).

Karst, O.J. Linear Curve Fitting Using Least Deviations. *J. Amer. Statist. Assoc.*, 1958, *53*, 118–132.

Kemperman, J.M.B. Least Absolute Value and Median Polish. (working paper, University of Rochester).

Khachian, L.G. A Polynomial Algorithm in Linear Programming. *Doklady Soviet Mathematics*, 1979, *20*, 191–194. (translated from Russian).

Klee, V. and Minty, G. How Good is the Simplex Method? In Shisha, O. (Ed.), *Inequalities*, New York: Academic Press, 1972.

Knuth, D.E. *The Art of Computer Programming*. Volume 3: *Sorting and Searching*. Reading, Mass.: Addison-Wesley 1975. (second printing).

Koenker, R. and Bassett, G. Regression Quantiles. *Econometrica*, 1978, *46*, 33–50.

Kohler, D.A. Translation of a Report by Fourier on His Work on Linear Inequalities. *Opsearch*, 1973, *10*, 38–42.

Krasker, W.S. and Welsch, R.E. Efficient Bounded-Influence Regression Estimation. *J. Amer. Statist. Assoc.*, 1982, *77*, 595–604.

Kuhn, H.W. and Quandt, R.E. An Experimental Study of the Simplex Method. *Proc. Symposia in Applied Math.*, 1963, *15*, 107–124.

Lai, T.L. and Robbins, H. Strong Consistency of Least Squares Esitmates in Regression Models. *Proc. Nat. Acad. Sci. USA*, 1977, *74*, 2667–2669.

Lai, T.L., Robbins, H., and Wei, C.Z. Strong Consistency of Least Squares Estimates in Multiple Regression. *Proc. Nat. Acad. Sci. USA*, 1978, *75*, 3034–3036.

Laplace, P.S. Sur Quelques Points du Systems du Monde.

Memoires de l'Academie Royale des Sciences de Paris, 1793, *Annee 1789*, 1–87. (reprinted in *Oeuvres Completes de Laplace, Vol. II*, Paris: Gauthier-Villars (1895)).

Laurent, P.J. *Approximation et Optimisation.* Paris: Hermann 1972.

Loeve, M. *Probability Theory.* Princeton, N.J.: Van Nostrand 1963. (3rd edition).

McCormick, G.F. and Sposito. V.A. Using the L_2 Estimator in L_1 Estimation. *SIAM J. Numer. Anal.*, 1976, *13*, 337–343.

Narula, S.C. and Wellington, J.F. Algorithm AS 108: Multiple Linear Regression with Minimum Sum of Absolute Errors. *Applied Statistics*, 1977a, *26*, 106–111.

Narula, S.C. and Wellington, J.F. Prediction, Linear Regression, and Minimum Sum of Relative Errors. *Technometrics*, 1977b, *19*, 185–190.

Narula, S.C. and Wellington, J.F. An Algorithm for the Minimum Sum of Weighted Absolute Errors Regression. *Commun. Statist. B*, 1977c, *6*, 341–352.

Nelson, P. A Note on Strong Consistency of Least Squares Estimates in Regression Models with Martingale Difference Errors. *Ann. Statist.*, 1980, *8*, 1057–1064.

Orveson, R.M. *Regression Parameter Estimation by Minimizing the Sum of Absolute Errors.* PhD thesis, Harvard University, 1969.

Osborne, M.R. and Watson, G.A. On an Algorithm for Discrete Nonlinear L_1 Approximation. *Comp. J.*, 1971, *14*, 184–188.

Pfaffenberger, R.C. and Dinkel, J.J. An Alternative to Least Squares. In David, H.A. (Ed.), *Contributions to Survey Sampling and Applied Statistics*, New York: Academic Press, 1978.

Quandt, R.E. and Kuhn, H.W. On Some Computer Experiments in Linear Programming. *Bulletin de l'Institute International de Statistique*, 1962, , 363–372.

Rabinowitz, P. Applications of Linear Programming to Numerical Analysis. *SIAM Review*, 1968, *10*, 121-159.

Rao, M.R. and Shrinivasan, V. A Note on Sharpe's Algorithm for Minimizing the Sum of Absolute Deviations in a Simple Regression Problem. *Management Sci.*, 1972, *19*, 222-225.

Reinsch, C.H. Smoothing by Spline Functions. *Numer. Math.*, 1967, *10*, 177-183.

Reinsch, C.H. Smoothing by Spline Functions, II. *Numer. Math.*, 1971, *16*, 451-454.

Rhodes, E.C. Reducing Observations by the Method of Minimum Deviations. *Phil. Mag.(Seventh Series)*, 1930, *9*, 974-992.

Robers, P.D. and Ben-Israel, A. An Interval Programming Algorithm for Discrete Linear L_1 Approximation. *J. Approx. Theory*, 1969, *2*, 323-336.

Robers, P.D. and Ben-Israel, A. A Suboptimization Method for Interval Linear Programming. *Linear Algebra and its* Applications, 1970, *3*, 383-405.

Robers, P.D. and Robers, S.S. Algorithm 458: Discrete Linear L_1 Approximation by Linear Programming. *CACM*, 1973, *16*, 629-631.

Rosenberg, B. and Carlson, D. A Simple Approximation to the Sampling Distribution of Least Absolute Residuals Regression Estimates. *Commun. Statist. B*, 1977, *6*, 421-437.

Ruppert, D. and Carroll, R.J. Trimmed Least Squares Estimation in the Linear Model. *J. Amer. Statist. Assoc.*, 1980, *75*, 828-838.

Sadovski, A.N. Algorithm AS74. L_1-Norm Fit of a Straight Line. *Applied Statistics*, 1974, *23*, 244-248.

Schlossmacher, E.J. An Iterative Technique for Absolute Deviations Curve Fitting. *J. Amer. Statist. Assoc.*, 1973, *68*, 857-865.

Schoenberg, I.J. Spline Functions and the Problem of Graduation. *Proc. Nat. Acad. Sci. USA*, 1964, *52*, 947–950.

Schor, N.Z. Cut-off Method With Space Extension in Convex Programming Problems. *Kibernetika*, 1977, *13*, 94–95. (translates as *Cybernetics* 13, pps. 94–96).

Seneta, E. On a Contribution of Cauchy to Linear Regression Theory. *Ann. de la Soc. Sci. de Bruxelles*, 1976, *90*, 229–235.

Seneta, E. The Weighted Median and Multiple Regression. *Austral. J. Statist.*, 1983, *to appear*, .

Seneta, E. and Steiger, W.L. A New LAD Curve Fitting Algorithm: Slightly Over-Determined Equation Systems in L_1. *Discrete Applied Math.*, 1983, *to appear*, .

Sharpe, W.F. Mean-Absolute-Deviation-Characteristic Lines for Securities and Portfolios. *Management Science*, 1971, *18*, B1–B13.

Sheynin, O.B. R. J. Boscovich's Work on Probability. *Archive for History of Exact Sciences*, 1973, *9*, 306–324.

Siegel, A.F. Low Median and Least Absolute Residual Analysis of Two-way Tables. *J. Amer. Statist. Assoc.*, 1983, *78*, 371–374.

Singleton, R.R. A Method for MInimizing the Sum of Absolute Values of Deviations. *Ann. Math. Statist.*, 1940, *11*, 301–310.

Sposito, V. Remarks on Algorithm AS74. *Applied Statistics*, 1976, *25*, 96–97.

Spyropoulos, K., Kiountouzis, E., and Young, A. Discrete Approximation in the L_1 Norm. *Computer Journal*, 1973, *16*, 180–186.

Steiger, W.L. *Linear Programming via Discrete L_1 Curve* Fitting. Technical Report 97, Department of Computer Science, Rutgers University, 1980.

Tukey, J.W. *Exploratory Data Analysis*. Reading, Mass.: Addison-Wesley 1977.

Turner, H.H. On Mr. Edgeworth's Method of Reducing Observations Relating to Several Quantities. *Phil. Mag. (Fifth Series)*, 1887, *24*, 466–470.

Usow, K.H. On L_1 Approximation II: Computation for Discrete Functions and Discretization Effects. *SIAM J. Numer. Anal.*, 1967, *4*, 233–244.

Utreras, F.I. On Computing Robust Splines and Applications. *SIAM J. Scientific and Statistical Computing*, 1981a, *2*, 153–163.

Utreras, F.I. Optimal Smoothing of Noisy Data Using Spline Functions. *SIAM J. Scientific and Statistical Computing*, 1981b, *2*, 349–362.

Wagner, H.M. Linear Programming Techniques for Regression Analysis. *J. Amer. Statist. Assoc.*, 1959, *54*, 206–212.

Wahba, G. Smoothing Noisy Data with Spline Functions. *Numer. Math.*, 1975, *24*, 383–393.

Witzgall, C. On Discrete L_1 Approximations. (working paper).

Yohai, V.J. and Maronna, R.A. Asymptotic Behavior of Least Squares Estimates for Autoregressive Processes with Infinite Variance. *Ann. Statist.*, 1977, *5*, 554–560.

Zeger, S.L. and Bloomfield, P. *Robust Smoothing Splines*. Technical Report 234, Department of Statistics, Princeton University, 1982. (Series 2).

Index of Names

Index of Subjects

PROGRESS IN PROBABILITY AND STATISTICS

Already published

PPS 1 Seminar on Stochastic Processes, 1981
E. Cinlar, K.L. Chung, R.K. Getoor, editors
ISBN 3-7643-3072-4, 248 pages, hardcover

PPS 2 Percolation Theory for Mathematicians
Harry Kesten
ISBN 3-7643-3107-0, 432 pages, hardcover

PPS 3 Branching Processes
S. Asmussen, H. Hering
ISBN 3-7643-3122-4, 472 pages, hardcover

PPS 4 Introduction to Stochastic Integration
K.L. Chung, R.J. Williams
ISBN 0-8176-3117-8
ISBN 3-7643-3117-8, 204 pages, hardcover

PPS 5 Seminar on Stochastic Processes, 1982
E. Cinlar, K.L. Chung, R.K. Getoor, editors
ISBN 0-8176-3131-3
ISBN 3-7643-3131-3, 310 pages, hardcover